SWITCHING AND TRAFFIC THEORY FOR INTEGRATED BROADBAND NETWORKS

THE KLUWER INTERNATIONAL SERIES IN ENGINEERING AND COMPUTER SCIENCE

COMMUNICATIONS AND INFORMATION THEORY

Consulting Editor

Robert Gallager

Other books in the series:

Digital Communications. Edward A. Lee, David G. Messerschmitt
  ISBN 0-89838-274-2

An Introduction to Cryptology. Henk C.A. van Tilborg
  ISBN 0-89838-271-8

Finite Fields for Computer Scientists and Engineers. Robert J. McEliece
  ISBN 0-89838-191-6

An Introduction to Error Correcting Codes With Applications. Scott A. Vanstone and Paul C. van Oorschot, ISBN 0-7923-9017-2

Source Coding Theory. Robert M. Gray
  ISBN 0-7923-9048-2

SWITCHING AND TRAFFIC THEORY FOR INTEGRATED BROADBAND NETWORKS

by

Joseph Y. Hui
Rutgers University

foreword by

Robert G. Gallager

KLUWER ACADEMIC PUBLISHERS
Boston/Dordrecht/London

Distributors for North America:
Kluwer Academic Publishers
101 Philip Drive
Assinippi Park
Norwell, Massachusetts 02061 USA

Distributors for all other countries:
Kluwer Academic Publishers Group
Distribution Centre
Post Office Box 322
3300 AH Dordrecht, THE NETHERLANDS

Library of Congress Cataloging-in-Publication Data

Hui, Joseph yu Ngai.
  Switching and traffic theory for integrated broadband networks / by Joseph Y. Hui.
    p. cm. — (The Kluwer international series in engineering and computer science. Communications and information theory)
  Includes bibliographical references.
  ISBN 0-7923-9061-X
  1. Telecommunication switching systems. 2. Broadband communication systems. I. Title. II. Series.
TK5103.8H85   1990
621.382—dc20   89-26704
                                                                        CIP

Copyright © 1990 by Kluwer Academic Publishers

All rights reserved. No part of this publication may be reproduced, stored in a retrieval system or transmitted in any form or by any means, mechanical, photocopying, recording, or otherwise, without the prior written permission of the publisher, Kluwer Academic Publishers, 101 Philip Drive, Assinippi Park, Norwell, Massachusetts 02061.

Printed in the United States of America.

CONTENTS

FOREWORD ix

PREFACE xi

Chapter 1. Integrated Broadband Services and Networks—
An Introduction 1

  1.1 Communication Networking 2
  1.2 Broadband Technologies 4
  1.3 Broadband Services 9
  1.4 To Integrate or Not to Integrate 11
  1.5 Outline of the Book 14
  1.6 Exercises 16
  1.7 References 21

PART I: SWITCHING THEORY

Chapter 2. Broadband Integrated Access and Multiplexing 25

  2.1 Time Division Multiplexing for Multi-Rate Services 26
  2.2 The Synchronous Transfer Mode 27
  2.3 The Asynchronous Transfer Mode 31
  2.4 Time Division Multiplexing Techniques for Bursty Services 35
  2.5 Switching Mechanisms by Time or Space Division 38
  2.6 Time Multiplexed Space Division Switching 42
  2.7 Appendix—What is in a Label? 47
  2.8 Exercises 48
  2.9 References 50

Chapter 3. Point-to-Point Multi-Stage Circuit Switching 53

  3.1 Point-to-Point Circuit Switching 53
  3.2 Cost Criteria for Switching 56
  3.3 Multi-Stage Switching Networks 60

3.4 Representing Connections by Paull's Matrix	61
3.5 Strict-Sense Non-Blocking Clos Networks	64
3.6 Rearrangeable Networks	66
3.7 Recursive Construction of Switching Networks	70
3.8 The Cantor Network	74
3.9 Control Algorithms	77
3.10 Exercises	80
3.11 References	82
Chapter 4. Multi-Point and Generalized Circuit Switching	**85**
4.1 Generalized Circuit Switching	85
4.2 Combinatorial Bounds on Crosspoint Complexity	89
4.3 Two Stage Factorization of Compact Superconcentrators	94
4.4 Superconcentrators and Distribution Networks—A Dual	99
4.5 Construction of Copy Networks	101
4.6 Construction of Multi-Point Networks	104
4.7 Alternative Multi-Point Networks	106
4.8 Exercises	108
4.9 References	111
Chapter 5. From Multi-Rate Circuit Switching to Fast Packet Switching	**113**
5.1 Depth-First-Search Circuit Hunting	114
5.2 Point-to-Point Interconnection of Multi-Slot TDM	117
5.3 Fast Packet Switching	119
5.4 Self-Routing Banyan Networks	126
5.5 Combinatorial Limitations of Banyan Networks	131
5.6 Appendix—Non-Blocking Conditions for Banyan Networks	134
5.7 Exercises	135
5.8 References	136
Chapter 6. Applying Sorting for Self-Routing and Non-Blocking Switches	**139**
6.1 Types of Blocking for a Packet Switch	140

Contents   vii

   6.2 Sorting Networks for Removing Internal Blocking   144
   6.3 Resolving Output Conflicts for Packet Switching   152
   6.4 Easing Head of Line Blocking   159
   6.5 A Postlude—Integrated Circuit and Packet Switching   161
   6.6 Appendix—Self-Routing Multi-Point Switching   164
   6.7 Exercises   170
   6.8 References   173

PART II: TRAFFIC THEORY

Chapter 7. Terminal and Aggregate Traffic   177
   7.1 Finite State Models for Terminals   178
   7.2 Modeling of State Transitions   182
   7.3 Steady State Probabilities   184
   7.4 Superposition of Traffic   186
   7.5 Traffic Distribution for Alternating State Processes 191
   7.6 Traffic Distribution for Poisson Processes   193
   7.7 Broadband Limits and The Law of Large Numbers   199
   7.8 Estimating the Traffic Tail Distribution   201
   7.9 Appendix—Improved Large Deviation Approximation   203
   7.10 Exercises   207
   7.11 References   208

Chapter 8. Blocking for Single-Stage Resource Sharing   211
   8.1 Sharing of Finite Resources   212
   8.2 Truncated Markov Chains and Blocking Probabilities   215
   8.3 Insensitivity of Blocking Probabilities   221
   8.4 The Equivalent Random Method   227
   8.5 Traffic Engineering for Multi-Rate Terminals   230
   8.6 Bandwidth Allocation for Bursty Calls   236
   8.7 Exercises   238
   8.8 References   242

Chapter 9. Blocking for Multi-Stage Resource Sharing   245
   9.1 The Multi-Commodity Resource Sharing Problem   246

9.2 Blocking for Unique Path Routing … 247
9.3 Alternative Path Routing—The Lee Method … 248
9.4 Assumptions for Approximating Blocking Probabilities … 254
9.5 Alternative Path Routing—The Jacobaeus Method … 256
9.6 Complexity of Asymptotically Non-Blocking Networks … 262
9.7 Exercises … 267
9.8 References … 269

Chapter 10. Queueing for Single-Stage Packet Networks … 271

10.1 The M/M/m Queue … 274
10.2 The M/G/1 Queue—Mean Value Analysis … 276
10.3 The M/G/1 Queue—Transform Method … 280
10.4 Decomposing the Multi-Queue/Multi-Server System … 284
10.5 HOL Effect for Packet Multiplexers … 290
10.6 HOL Effect for Packet Switches … 291
10.7 Load Imbalance for Single-Stage Packet Switches … 298
10.8 Queueing for Multi-Cast Packet Switches … 301
10.9 Exercises … 304
10.10 References … 310

Chapter 11. Queueing for Multi-Stage Packet Networks … 313

11.1 Multi-Stages of M/M/1 Queues … 314
11.2 Open and Closed Queueing Networks … 322
11.3 Application to Multi-Stage Packet Switching … 326
11.4 Analysis of Banyan Network with Limited Buffering … 331
11.5 Local Congestion in Banyan Network … 334
11.6 Appendix—Traffic Distribution for Permutations … 336
11.7 Exercises … 338
11.8 References … 340

INDEX … 343

Foreword

The rapid development of optical fiber transmission technology has created the possibility for constructing digital networks that are as ubiquitous as the current voice network but which can carry video, voice, and data in massive quantities. How and when such networks will evolve, who will pay for them, and what new applications will use them is anyone's guess. There appears to be no doubt, however, that the trend in telecommunication networks is toward far greater transmission speeds and toward greater heterogeneity in the requirements of different applications.

This book treats some of the central problems involved in these networks of the future. First, how does one switch data at speeds orders of magnitude faster than that of existing networks? This problem has roots in both classical switching for telephony and in switching for packet networks. There are a number of new twists here, however. The first is that the high speeds necessitate the use of highly parallel processing and place a high premium on computational simplicity. The second is that the required data speeds and allowable delays of different applications differ by many orders of magnitude. The third is that it might be desirable to support both point to point applications and also applications involving broadcast from one source to a large set of destinations.

The second major problem is that of traffic analysis or performance analysis for future networks. The roots for this problem again go back to telephony and to data networks, both of which are rooted in queueing theory. The new twists here are, first, that studying traffic flowing through various stages of the switches is more important than studying the traffic on the network links. Next, the heterogeneity of required speeds and allowable delays for different applications is important, and also the possibility of broadcast applications is important.

The reader will be pleasantly surprised to find that most of the topics here are quite accessible. An undergraduate senior in electrical engineering or computer science certainly has the background (although not necessarily the maturity) to understand switching for future networks as developed in the first half of the book. The traffic analysis in the second half requires quite a bit of facility with probability, but given this facility, the material is again surprisingly accessible.

Many of the results here are not found elsewhere in book form and have only appeared relatively recently in journals. Professor Hui, however, has managed to pull all these results together into an integrated whole. There is quite a bit of discussion about engineering constraints and practical aspects of the subject, but it is placed within a conceptual framework that helps the reader develop an overall understanding and intuition for the issues involved in designing high speed networks. There will probably be a large number of books coming out in this area in the next few years, and one hopes that this book will set a standard that these other books will follow.

Robert G. Gallager
Fujitsu Professor of Electrical Engineering
Co-Director, Laboratory for Information and Decision Systems
Massachusetts Institute of Technology

Preface

This is a senior or graduate level textbook intended for students and researchers in three different fields. First and primarily, this book is intended for those in the field of futuristic telecommunication networks. These networks, yet to be implemented and understood completely, provide flexible point-to-point and multi-point communications via media such as voice, data, and video. The economy of these networks relies on the recent advances in broadband transmission and processing technologies. The time has come for us to reexamine the well-established switching and traffic theory for telephone service to heterogeneous services with diverse attributes. This book attempts to frame a modern view of the theory based on the rich traditions of classical telephony.

Second, this book is intended for those in the field of parallel computation. The fields of telecommunication switching and parallel computing have traditionally been using methodologies from the other field, enriching both fields in the process. This cross-fertilization is not surprising since communication networks can be used for distributed processing, whereas computer networks can be used for exchange of information. Readers interested in parallel computing should find the switching networks described applicable for the interconnection of a large number of processors. Similarly, teletraffic theory can be applied to computer network and parallel processing performance analysis.

Third, this book is intended for those in the field of optical transmission and processing technologies. For making good use of these new technologies, it is helpful to understand the structural, functional, and control attributes of switching networks. These attributes in turn determine how well heterogeneous services may be supported. The relative strength and weakness of these technologies should be evaluated with respect to these attributes. Therefore, the proper choice of technologies and their deployment configuration in a communication network affect the variety and quality of services. This book is intended to furnish the necessary background for the optical technologist to build communication subsystems from optical components.

This book was developed from course notes for a class taught at the Center for Telecommunications Research at Columbia University. In choosing the topics to cover, the author became keenly aware that a vast but scattered literature exists for classical switching and traffic theory, whereas research in the field of integrated broadband switching is in a fledgling stage. In fact, the feasibility of an integrated network remains a thesis to be proven since none has been implemented on a large scale yet. Some of the criteria for feasibility are yet to be defined. Nevertheless, this book attempts to instill a set of principles which may prove useful for constructing and analyzing integrated broadband networks. No attempt is made for completeness since the subjects covered remain a small subset of the available literature. Yet hopefully, the material should define an essential framework for the field. Most of the material is chosen for its theoretical import, and some sections reflect the research interests of the author as well as recent research activities at Bellcore. Many sections present new results which have not been published before.

A first course in Probability is necessary for understanding the material presented. A prior knowledge of communication networks and computer algorithms is helpful but not crucial. To facilitate easier understanding, new and simple proofs are given for most results, supplemented extensively with intuition and illustrations. The material can be taught in one semester, or more leisurely divided into two courses on switching and traffic respectively, with the supplement of recent research papers and the requirement of a term project. A subset of this material may be chosen for a short course for research engineers in the communication industry. A teacher's manual is also available.

Preface

As the title *Switching and Traffic Theory for Integrated Broadband Networks* suggests, the book is divided into two parts: *Switching Theory* (Chapters 2-6), and *Traffic Theory* (Chapters 7-11). Chapter 1 serves as a prelude by introducing the technology, system, and service aspects of an integrated broadband network. There is a rough correspondence between chapters in the two parts. This correspondence, together with a description of the chapters, is given at the end of chapter 1.

This book, as well as my research contributions described therein, is made possible by the gracious and enriching environment of several institutions. Foremost, I would like to thank Bell Communications Research and its management (Eric Nussbaum, Dave Sincoskie, Steve Weinstein, and Pat White) for their generous support during the past years when most of the research and writing were done. I am fortunate to be at this research institution when and where many seeds of the so called broadband revolution fermented. I would also like to thank the Center for Telecommunications Research of Columbia University and its faculty (Aurel Lazar, Henry Meadows, and Mischa Schwartz) for fruitful interactions, and for allowing me to teach the material. Many individuals reviewed the material, including Eric Addeo, Howard Bussey, Fred Descloux, Bob Gallager, Alex Gelman, Shlomo Halfin, Mike Honig, Ram Krishnan, Francois Labourdette, Tony Lee, Bob Li, Victor Mak, Steve Weinstein, Albert Wong, Liang Wu, and two anonymous individuals. Their criticism improved the presentation significantly as well as weeded out many errors. Finally, I want to thank my wife Ruth for her encouragement and care over many weekends and week nights I spent typing away at the terminal, and my little Justin for roaming cheerfully around me in his walker during those times. Without them, life would seem strangely disconnected.

Joseph Y. Hui
Piscataway, NJ.
July, 1989.

SWITCHING AND TRAFFIC THEORY FOR INTEGRATED BROADBAND NETWORKS

CHAPTER 1

Integrated Broadband Services and Networks - An Introduction

- *1.1 Communication Networking*
- *1.2 Broadband Technologies*
- *1.3 Broadband Services*
- *1.4 To Integrate or Not to Integrate*
- *1.5 Outline of the Book*

In recent years, significant advances have been made in the areas of communication and computation technologies. The advances in communication technologies are mostly in the area of transmission and switching devices. These device technologies have substantially outpaced their applications in building communication networks and providing new communication services. In this chapter, we first review some key technologies which will drive the advancement of communication networking. We also examine how the proliferation of new communication services requires an integrated approach to communication networking.

We tag these technologies and services as *broadband* in the sense that their speed ranges from 1 Mb/s (10^6 bits per second) to 100 Mb/s and beyond.

1.1 Communication Networking

Before we describe how these broadband technologies may be deployed for communication networking, we take a brief look at the history of telephone networking.

It is often said that Alexander Graham Bell did not invent only a telephone, but two telephones; hence the issue of interconnecting these telephones was crucial right from the beginning of the telephone industry. One method of interconnection provides a direct wire connection between any two phones (figure 1). Hence a total of $n(n-1)/2$ wires are required to interconnect n telephones. Since each telephone has $n-1$ incident wires, a selector is needed at each telephone to connect the telephone to one wire at a time.

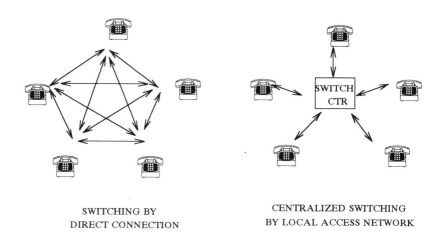

SWITCHING BY
DIRECT CONNECTION

CENTRALIZED SWITCHING
BY LOCAL ACCESS NETWORK

Figure 1. The Local Access Network

The above solution is obviously inefficient. An alternative solution brings the other end of the telephone wire connected to a telephone to a switching center, where telephone operators provide manual connection between pairs of telephones. Thus only n wires are needed

1.1 Communication Networking

to interconnect n telephones. However, a switching mechanism to interconnect the n wires incident to the switching center is needed. We call this network the local access network, which is also shown in figure 1.

To facilitate communications between areas served by different switching centers, wire connections called trunks are used between pairs of switching centers. The grouping of trunks between 2 switching centers is called a trunk group. To interconnect n switching centers, we may need at least $n(n-1)/2$ trunk groups. In practice, it may not be convenient to provide a trunk group between every pair of switching centers. Without a direct trunk group between 2 switching centers, a telephone connection may have to be made through other switching centers as intermediates as shown in figure 2. We call this network between the local switching centers the exchange network.

For calls traversing even longer distances, each switching center is connected to a long distance switching center which routes the call to other long distance switching centers or local switching centers (figure 3). A direct long distance trunk group may connect the local switching center to the long distance center, or the connection may be made via the exchange network. We call this network the long distance network.

Hence a hierarchical network is created. This hierarchy is created for the purpose of concentrating traffic from the lower levels of the hierarchy to the higher levels. This concentration reduces the transmission cost for making a connection between two distant telephones at the expense of increased switching cost in the hierarchy of switching centers. Networking therefore involves a tradeoff between transmission cost and switching cost at each level of the network hierarchy. Without a local switching center, every telephone would require a direct link to every other telephone. Hence the total number of transmission links grows quadratically with the number of telephones. Given that telephones are switched by local switching centers, there is a tradeoff between the size of the area served by a switching center versus the total number of switching centers. Having small centers would require more centers to serve a fixed size area, but the switches and local access wires would be less costly. Again the total number of pairs of switching centers grows quadratically as the number of switching centers, resulting in increased trunking costs. This trunking cost can be reduced by employing a long distance switching center one level up the network hierarchy, with reduced number of direct links between far apart local switching centers but

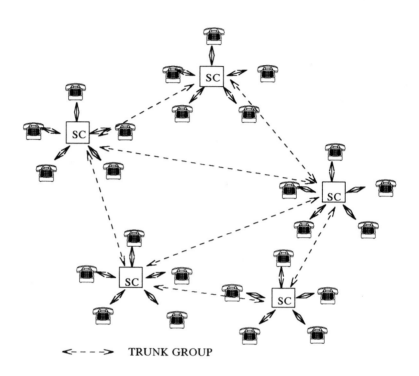

Figure 2. The Exchange Network

requiring large capacity switches for routing long distance traffic.

This tradeoff between transmission costs and switching costs depends strongly on the technologies for building transmission and switching facilities. We shall examine how some key technologies developed, focusing on their influence on network structure and on the tradeoff of transmission versus switching.

1.2 Broadband Technologies

The step-by-step mechanical switch was invented by Strowger to replace manual attendant switching. Mechanical means of providing interconnections were in turn replaced by electronic crosspoint switches. With the evolution of larger mechanical and electronic

1.2 Broadband Technologies

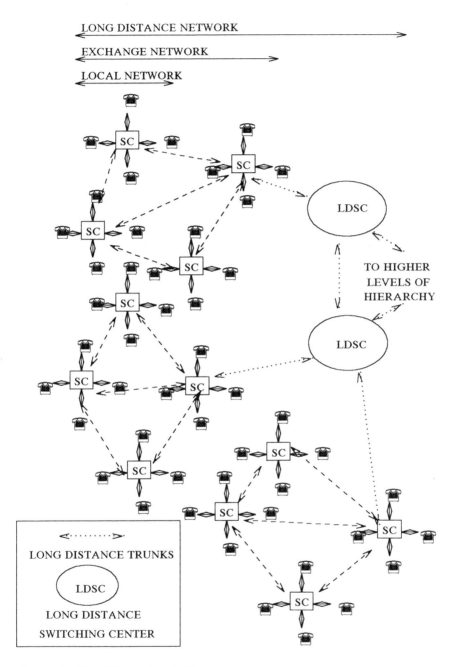

Figure 3. The Hierarchical Network

switching systems, it became necessary for one to understand the mathematical theory of interconnection networks. Research in this area produced a fundamental understanding of circuit switching networks constructed by interconnecting many smaller switching elements in stages.

Gradually, stored program computing was used to provide efficient control and utilization of the switching network. Stored program control also provided flexible management of the switching systems, and offered diverse features for telephone services. With the adoption of multi-stage switching and stored program control, large electronic switching systems for local and long distance switching emerged.

The earlier switching machines used in the network were space division switches, in which a telephone connection has exclusive use of a physical path in the switching network for the entire duration of the connection. The switching mechanism is based on either mechanically or electronically toggling connections among electrical conductors.

Almost in parallel, long distance transmission technology matured with the development of reliable analog amplifiers as well as microwave and coaxial transmission systems. Thus, analog voice signals were multiplexed via Frequency Division Multiplexing (FDM) on these high capacity transmission media, with each signal using a frequency slot in the frequency spectrum. Once multiplexed, these signals could be amplified and transmitted at low cost over long distances. However, the relatively high cost of multiplexing made these wide band transmission systems suitable only for long distance and high volume routes for which the reduction in transmission cost more than compensated for the increase in multiplexing cost. Thus, the deployment of frequency division multiplexing consolidated the long distance network into fewer routes with higher capacity.

The foundations of broadband communication were laid with the development of digital technology, which samples a continuous time signal at discrete instants and represents the sampled value in digital formats. These digital sampled values are often represented by binary numbers. Given this representation, the bit streams of different signals can be multiplexed in time slots to form a single stream for transmission on a communication channel. This form of sharing of a transmission medium is known as Time Division Multiplexing (TDM), which we shall explain in more detail in the next chapter. Digital communication in the telephone network started with deploying digital

1.2 Broadband Technologies

transmission lines multiplexing tens of telephone connections between switching centers.

With lowered cost and increased speed of digital electronics, the multiplexing cost of time division multiplexing became less costly than that for frequency division multiplexing. Thus, time division multiplexing was gradually used for the local access network and the exchange network, while analog frequency multiplexing still dominated the long distance network due to its lower transmission cost. In the local access network, the application of time multiplexed digital transmission also enabled several telephones and data terminals to access the local switching center via a single wire.

In fact, time division multiplexing is more than just a multiplexing and transmission mechanism; it can also be used as a switching mechanism. Terminals or interfaces distributed along a shared transmission line may time multiplex their information transmission, and select the information from the transmission line destined for that terminal or interface. Therefore, "switched" connections are achieved for terminals served by the same transmission line. Consequently, the distinction between transmission and switching became less obvious in these network architectures.

Besides this distributed switching mode, time division switching can also be applied to building switching machines, which predominantly used space division switching before time division switching was introduced. Suppose the inputs and outputs of a space division switch are transmission lines multiplexed by the same slot timings. By changing the connections of the space division switch per time slot according to the input-output connection pattern of that time slot, we have a time multiplexed space division switch. We shall describe this mechanism in more detail in the next chapter. For time multiplexed space division switches, the traffic volume which may be handled increases linearly with the speed of the switching network.

Of course, there are other benefits associated with digital communications, such as reliable regeneration of digital signals between transmissions, and ease of digital signal processing. Further, data terminals communicate nothing other than binary bit streams, and hence the ideal network for such terminals should be all digital. The telephone network was originally designed for analog telephony, consequently it is ill-suited for data communication. As a result, a new class of networks designed distinctly for data communication emerged, and many data networks either overlay or bypass the telephone network.

Transmission speed increased exponentially in the eighties with the development of fiber optics and laser technologies. Reliable transmission system are now deployed at Gb/s (10^9 bits per second) rates and may be increased to tens or even hundreds of Gb/s with Gallium Arsenide devices, picosecond optics, coherent lightwave detection, wavelength division multiplexing, and possibly high temperature superconducting. This continuing technological development causes a substantial disparity between transmission bit-rate and terminal bit-rate. Fiber optics is more reliable and economical for long distance transmission than analog microwave transmission. For the exchange network, this extreme economy of shared transmission may simplify the network structure. Since the aggregate telephone traffic between most local switching centers may very well be less than the capacity of a single fiber, it may be more economical to consolidate the exchange network into a hubbing architecture, which connects local switching centers in an exchange network to a hub switching center using high capacity optical transmission lines. Traffic between any two local switching centers is switched through the hubbing center instead of through a direct line between the local switching centers, or through other intermediate switching centers in tandem. Consequently, a large capacity switch is needed at the hubbing center.

It is inevitable that these new transmission technologies are component-wise more expensive than more mature technologies such as the copper wire, despite their dramatic cost advantage when shared by many calls. Thus replacing the local access network on a component for component basis using these new technologies often may not be economical unless we utilize new network architectures or new services that capitalize on the increased bandwidth capability. Consequently, the introduction of high bandwidth services would also increase many-fold the traffic volume at higher levels of the network hierarchy, thereby affecting the structure of the exchange network and the long distance network.

In summary, we have seen how the development of technologies causes shifts in the relative use of various switching and multiplexing techniques, namely space division, time division, and frequency division. The future network is likely to make use of not just one but all three techniques. We may have an optical fiber transporting many time multiplexed bit streams over different wavelengths of light. Optical space division switches may multiplex these signals onto

1.2 Broadband Technologies

different fibers. We shall have a more detail description of these multiplexing techniques in chapter 2.

With the introduction of optical transmission into the local access network comes the introduction of many broadband services that may not be possible with the current copper network. These services will be briefly described in the next section.

1.3 Broadband Services

Our society is becoming more informationally and visually oriented. Personal computing facilitates easy access, manipulation, storage, and exchange of information. These processes require reliable transmission of data information. Communicating documents by images and the use of high resolution graphics terminals provide a more natural and informative mode of human interaction than just voice and data. Video teleconferencing enhances group interaction at a distance. High definition entertainment video improves the quality of picture at the expense of higher transmission bit-rates, which may require new transmission means other than the present overcrowded radio spectrum.

These new communication services depart from the conventional telephone service in three essential aspects. They can be multi-media, multi-point, and multi-rate. In contrast, conventional telephony communicates using the voice medium only, connects only two telephones per call, and uses circuits of fixed bit-rates. These differences force us to examine multiplexing and switching issues in a radically different way. We shall examine each difference individually.

A multi-media call may communicate audio, data, still images, or full-motion video, or any combination of these media. Each medium has different demands for communication qualities, such as bandwidth requirement, signal latency within the network, and signal fidelity upon delivery by the network. Moreover, the information content of each medium may affect the information generated by other media. For example, voice could be transcribed into data via voice recognition and data commands may control the way voice and video are presented. These interactions most often occur at the communication terminals but may also occur within the network.

A multi-point call involves the setup of connections among more than two people. These connections can be multi-media. They can be one way or two way connections. These connections may be reconfigured many times within the duration of a call. Let us give a few examples here, to contrast point-to-point communications versus multi-point communications. Traditional voice calls are predominantly two party calls, requiring a point-to-point connection using only the voice medium. To access pictorial information in a remote database would require a point-to-point connection that sends low bit-rate queries to the database, and high bit-rate video from the database. Entertainment video applications are largely point-to-multi-point connections, requiring one way communication of full motion video and audio from the program source to the viewers. Video teleconferencing involves connections among many parties, communicating voice, video, as well as data. Thus offering future services may require more than providing just the transmission and switching for making a connection, but also the flexible management of the connection and media requests of a multi-point, multi-media communication call.

A multi-rate service network is one which allocates transmission capacity flexibly to connections. A multi-media network has to support a broad range of bit-rates demanded by connections, not only because there are many communication media, but also because a communication medium may be encoded by algorithms with different bit-rates. For example, audio signals can be encoded with bit-rates ranging from less than 1 kb/s to hundreds of kb/s, using different encoding algorithms with a wide range of complexity and quality of audio reproduction. Similarly, full motion video signals may be encoded with bit-rates ranging from less than 1 Mb/s to hundreds of Mb/s. Thus a network transporting both video and audio signals may have to integrate traffic with a very broad range of bit-rates.

Besides a broad range of bit-rates from different terminals, each terminal may have a time varying bit-rate requirement over the duration of a connection. Different terminals may exhibit a varying amount of burstiness. To give an idea of the range of burstiness, we consider a rough measure defined by the ratio of maximum instantaneous bit-rate within the duration of a connection to the average bit-rate over the duration of the connection. This burstiness may range from 2 to 3 for speech encoders which prevent transmission during silent intervals. The degree of burstiness can be higher for applications such as computer communications and on-line database queries.

1.3 Broadband Services

Our focus throughout this book is on understanding multiplexing, switching, and performance issues for establishing multi-point, multi-media and multi-rate connections in an integrated services network. Before doing that, we need a more careful examination of the notion of network integration.

1.4 To Integrate or Not to Integrate

Traditionally, the services mentioned before are carried via separate networks - voice on the telephone network, data on computer networks or local area networks, video teleconferencing on private corporate networks, and television on broadcast radio or cable networks. These networks are largely engineered for a specific application and are ill-suited for other applications. For example, the traditional telephone network is too noisy and inefficient for bursty data communication. On the other hand, data networks which store and forward messages using computers have very limited connectivity, usually do not have sufficient bandwidth for digitized voice and video signals, and suffer from unacceptable delays for these real time signals. Television networks using the radio or the cable medium are largely broadcast networks with minimal switching facilities.

It is often desirable to have a single network for providing all these communication services in order to achieve the economy of sharing. This economy motivates the general ideal of an integrated services network. Integration avoids the need for many overlaying networks, which complicate network management and reduce the flexibility in the introduction and evolution of services. This integration is made possible with the advances in broadband technologies and high speed information processing.

However, integration within the network can have different meanings, depending on the part and the function of the network being considered. We shall examine these more specific notions of integration shown in figure 4.

First, integrated access involves the sharing, among services from an end user, of a single interface to a single transmission medium in the local access network. A well integrated access network should provide flexible multiplexing of as many services as possible. For broadband services, optical fiber is the transmission medium of choice for all these applications, due to its excellent transmission reliability and high bandwidth. Given the abundant bandwidth of optical

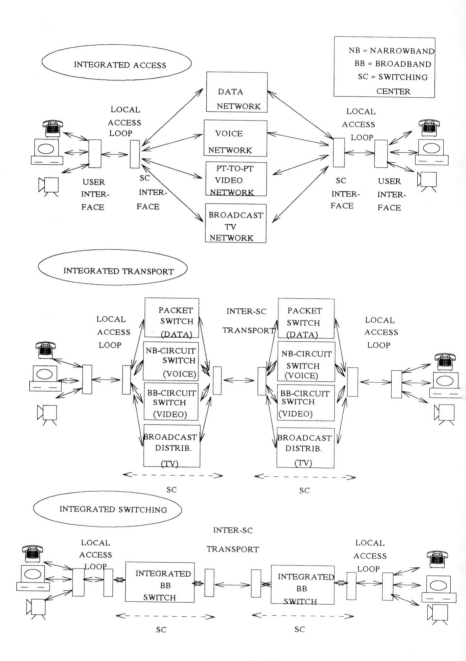

Figure 4. Levels of Integration of Networks

1.4 To Integrate or Not to Integrate

transmission, there remains the question of how the broad range of service bit-rates can be flexibly multiplexed through a single interface onto the transmission medium. We shall focus on this issue in the next chapter.

Second, integrated transport involves the flexible sharing, among services from possibly many users, of high capacity transmission links in the network. Integrated transport avoids the segregation of different traffic types and media onto different transmission links, hence may facilitate easier interactions between media within the network.

Third, integrated switching involves switching multi-rate, multi-media services within a single switching machine, in particular a single interconnection network. An integrated switch would avoid the necessity of adding a new interconnection network whenever a new service of distinct traffic characteristic is introduced. An integrated switching network must be flexible enough to meet the delay and bit-rate requirements of each service.

Fourth, integrated call processing involves the sharing of communication software for calls of different multi-media, multi-rate, and multi-point configurations. Integrated call processing provides a uniform and flexible functional description of calls, and uses a single procedure to map the resource requirements of calls onto the available physical resources in the network. A flexible representation of calls and network resources has many advantages. It provides flexible design and deployment of new services. It simplifies network management and maintenance. Furthermore, an integrated logical view of calls decoupled from the physical view of the network allows the call processing software to be independent of switch hardware, thereby allowing the reuse of software for a variety of evolving hardware.

Despite the above advantages of integration, there are practical reasons why the process of integration can be slow and at times impossible in the public network. Since the building and upgrading of a network is a slow process, the integration and upgrading ideally should be compatible with the existing network. In other words, we may replace a terminal or part of a network in the process, but such a replacement must not require replacing all other terminals or parts of the network. For example, the replacement of trunk wires between local switching centers with digital time division multiplexed carriers is compatible with and transparent to an analog network, as is replacing space division analog switches with time division digital switches.

When we introduce optical technologies and multi-rate multiplexing methods into the network, we face the similar problem of compatibility between a proposed broadband transport architecture and the existing single rate and fixed delay telephone network.

Network evolution and integration can be a long and difficult process. The first and probably the easiest step towards network integration is access integration, the subject of the next chapter.

1.5 Outline of the Book

As the title of the book *Switching and Traffic Theory for Integrated Broadband Networks* suggests, the book is roughly divided into two halves: *Switching Theory* and *Traffic Theory*. This division is due to the different disciplines used; *Switching Theory* is largely combinatorial and algorithmic, while *Traffic Theory* is largely statistical. We shall give several theoretical results which may prove useful for the design and analysis of switching systems.

As an introduction, we have discussed the technology and service aspects of broadband networking in chapter 1. We then launch into some of the more well-known results of *Switching Theory* in the next five chapters. The stage for the discussion of this theory is set in chapter 2 which deals mostly with methods of multiplexing for integrated networks. They include rigid bandwidth allocation methods termed circuit switching, and more flexible but variable delay bandwidth allocation methods termed packet switching. Towards the end of the chapter, we describe the implementation of time domain switching and space domain switching, and show that mathematically there is a certain equivalence between these two domains. Later, we show how space and time domain switching can be simultaneously used in a form called Time Multiplexed Switching (TMS). TMS is one example of multi-stage switching networks, the central theme of switching network theory.

Chapter 3 presents the classical circuit switching theory initiated by Clos. The associated switching networks make point-to-point connections with fairly optimal hardware complexity but requires a high degree of control complexity. This classical theory has been employed in most modern electronic telephone switching machines.

Chapter 4 generalizes connectivity from point-to-point to multi-point connections as well as concentration. We first derive some

1.5 Outline of the Book

combinatorial bounds for these generalized connection functions, and then give some practical constructions. In the process, we point out the intimate relation between concentration, distribution, and multi-casting.

Chapter 5 generalizes connectivity from single-rate to multi-rate connections. We first look at some problems arising from multi-slot circuit switching. We then introduce the concept of self-routing fast packet switching to simplify control through statistical switching. The same multi-stage packet switching technique has been implemented in parallel computing systems as a communication means among processors.

Chapter 6 describes self-routing and internally non-blocking switching networks employing sorting as a switching mechanism. The same sorting mechanism can make these networks function as a packet switch. These networks speed up network control at the expense of more hardware.

The next five chapters, namely chapter 7-11, will present some analytical tools of *Traffic Theory*. The reader readily realizes a loose parallel with the earlier chapters - they evaluate somewhat in sequence the blocking and throughput performance of switching networks in the earlier chapters.

Chapter 7 sets the stage for blocking, throughput and delay analysis in later chapters by first of all modeling the bit-rate processes from communication terminals, and then analyzing the statistical properties of aggregate traffic obtained by superposing the traffic of individual terminals. This chapter can be viewed also as a review of well-known stochastic processes and methods of probability analysis.

Chapter 8 analyzes the blocking probability for a single stage of shared network resources for a given aggregate traffic. Particularly, we focus on circuit multiplexing, which is described in chapter 2.

Chapter 9 extends the blocking analysis of chapter 8 to the multi-stage circuit switching networks. Therefore chapter 9 is complementary to chapter 3, which introduces multi-stage interconnection networks for circuit switching. Unfortunately, we do not have yet a satisfactory traffic theory for the multi-cast switching system which is introduced in chapter 4.

Chapter 10 analyzes the delay and throughput performance for a single stage queueing system. The first half is devoted to several useful results in queueing theory. The second half extends the theory to the analysis of internally non-blocking packet multiplexers and switches. The analysis is complementary to chapter 6, which introduces sorting as a switching mechanism for internally non-blocking packet switching.

Chapter 11 extends the queueing analysis of chapter 10 to multi-stage queueing networks. Such analysis gives an end-to-end delay and congestion description for networks of packet switching systems. On the other hand, such analysis can also be applied to internally buffered multi-stage interconnection networks, such as the fast packet switching network introduced in chapter 5.

Instead of following this structure/performance dichotomy dictated by the chapter sequence, it may also be advantageous to read the corresponding chapter on performance right after a switching structure is introduced. This may perhaps give a better comparison between the switching structures of the earlier chapters. We shall provide pointers as necessary within some sections in each chapter.

While performance analysis provides a mean of comparison among switching network architecture, one should avoid the temptation of making broad statements concerning the superiority of one switch design over another. Very often, the choice of a switch design depends on the application and network context, reliability and operational issues, as well as a myriad of human, social, and economics factors. Despite the lack of a quantitative discipline for a tradeoff of these matters, I hope the reader may nevertheless be intrigued by the beauty of the science of interconnection. The switching structures chosen for description in this book are by no means exhaustive. Certainly, much more remains to be discovered in this rather new discipline.

1.6 Exercises

One important aspect in understanding telecommunication networking is appreciating the scales of the network being studied. In this exercise, we shall first cite certain statistics for the Public Switched Telephone Network (PSTN) in the United States, and then ask the readers to analyze these statistics. These statistics taken from reference [9] reflect the PSTN in the United States (including the parts managed the Bell System and by the independent telephone companies) before the 1984 divestiture of the Bell System.

1.6 Exercises

There are 180 million telephones in the U.S. The Bell System provides access to 80% of these telephones versus 20% for the independent companies. The rate of making calls for a telephone fluctuates in time (from day to day and from hour to hour) as well as in location (urban, suburban and rural).

Telephone switching centers in the U.S. are largely hierarchical. There are five classes of switching centers, with class 1 being the highest regional centers and class 5 being the lowest local access centers. The numbers of centers from class 1 to 5 are 10, 52, 168, 933, and 18803 respectively.

Telephone calls in the U.S. are classified as local calls or long distance calls. Local calls are routed by class 5 switching centers in tandem through the exchange network, whereas long distance calls are routed by class 4 or higher level centers through the long distance network. 246 billion local calls are placed annually, versus 32.5 billion long distance calls placed annually.

Let us now focus on the local access network. As mentioned previously, we shall classify the local switching centers into three types: urban, suburban, and rural. These three types of centers have different average parameters which we cite as follows.

	urban	**suburban**	**rural**
Areas served (square miles)	12	110	130
Local access lines (thousands)	41	11	0.7
No. of switching systems	2.3	1.3	1.0
Intraoffice calling percentage	31	54	66
No. of trunks	5000	700	35
No. of trunk groups	600	100	5.5
Hundred call seconds/hour	3.1	2.7	2.1

Some explanations of the above terms are given here. Each center serves as a termination for the local access lines in the areas served.

Roughly 95% of the centers serve less than 100 thousand lines, and the really large centers serve somewhere between 200 to 300 thousand local access lines. Each center may have more than one switching systems. Each system is also termed a central office in Bell System terminology. The central offices are connected by trunks organized into trunk groups. The intraoffice calling percentage is the percentage of calls switched by one switching system only. The number of hundred call seconds (CCS) per hour is the hundreds of seconds the telephone is used during the busy hour.

Traffic is measured in terms of erlang, after the founder of traffic theory A. K. Erlang. Let us define this measure here since we shall be using it in the exercise. An erlang is a dimensionless unit of traffic intensity used to express the average number of calls underway or the average number of devices in use [9, pp. 794]. For example, a trunk group with 20 trunks may be measured to carry a traffic of 10 erlangs averaged say over an hour, meaning a 50% utilization of the trunks in an hour. We may also use this term to measure the capacity of facilities. For example, the 20 trunks theoretically have a capacity of 20 erlangs. However, they may typically have only 15 erlangs of capacity in practice. To allow it to carry more than the practical capacity may cause excessive call blocking, which shall be defined and studied in chapter 8.

1. Consider the service area of a local switching center.
    a. From the above figures, estimate the population density for the area served by each of the three types of local switching centers. You may assume a telephone per capita.
    b. Now estimate the population density at where you live, and get a picture of the kind of switching center serving you. Also, estimate the percentage of your phone calls made to areas outside that served by your local switching center, in comparison to the intraoffice call percentage quoted above.
2. Consider the pattern of calling rate in a day.
    a. Using the annual figures given, compute the rate of calling (calls/hour/phone) averaged over a year.
    b. Compute the rate of calling for the busy hour (typically during late morning), using the CCS figures for the three types of switching centers and assuming that a typical call lasts for 3 minutes.

1.6 Exercises

   c. What can you say about the unevenness of calling rate?

   d. Compare your calling habit (frequency and duration of calls) between the busy hour and the rest of the day.

3. Consider the aggregate of telephone traffic in terms of bits per second and assuming that a telephone call requires a 64 kb/s transport.

   a. Estimate the total bit rate for the entire U.S. during the busy hour when each phone is offering a traffic of 0.1 erlang.

   b. Estimate the total bit rate on the local access lines for the three types of switching centers during the busy hour.

   c. Repeat parts a and b with these telephones replaced by picture phones. (Assume 1.5 Mb/s for conference grade video and approximately 100 Mb/s for broadcast quality video. [9, pp. 528].)

4. Assume each optical fiber has a transmission capacity of 2.4 Gb/s.

   a. During the busy hour, how many fibers are required to carry the telephone traffic for the entire U.S. and for each of the three types of local switching centers?

   b. How many fibers are required to carry picture phone traffic for the entire U.S. and for each of the three types of local switching centers?

   c. This book can be stored in a 2 Megabyte memory. Estimate how long it takes to transmit all books in a local library with ten thousand similar volumes on a fiber transmission link?

5. Consider how the intraoffice calling percentage affects the size of the switching system. Specifically, consider L local access lines terminating at a switching system, with a traffic of a erlang per line. Let us assume an intraoffice calling percentage of R %.

   a. How many trunks would be needed for the switching system, assuming each trunk has a practical capacity of $b < 1$ erlang?

   b. What is the size of the switching system (namely the number of access lines and trunk lines assuming each line can carry a two-way conversation)?

c. What is the total amount of traffic switched measured in erlang?

6. (Continuation)

   a. Using the figures of hundred calling seconds (CCS), compute a for the busy hour for the three types of local switching centers.

   b. Assuming $b=a$, compute the size of the switching system for the three types of local switching centers.

   c. What is the total traffic switched by each switching system?

7. Consider the load on trunks between central offices.

   a. Using the number of trunks per central office as well as the previously computed trunk traffic, compute the traffic carried per trunk from the three types of switching centers.

   b. Compute also the size of the trunk group. Comment on the practical capacity of a trunk versus the size of the trunk group containing the trunk.

8. Consider the traffic at various level of the network hierarchy for long distance telephone traffic. For lack of better statistics, assume a tree structure for the network, and also that the traffic is shared evenly among the given number of switching centers in each class. For a tree structure, a long distance call has a unique path up to a certain level of the hierarchy before it is routed down the hierarchy. Furthermore, assume that $1/2$ of the calls are routed beyond a class 4 center, $1/4$ are routed beyond a class 3 center, $1/8$ are routed beyond a class 2 center, and $1/16$ have to be routed from a class 1 center to another class 1 center before being routed down the hierarchy. Assume that 32 billion long distance calls are placed per year, and that a rate of 1 billion three-minute calls per year is roughly equivalent to a traffic of 5 thousand erlangs.

   a. Compute the traffic switched at each center at different levels.

   b. Compute also the number of switch centers each center is connected to, and the traffic volume on each connection.

c. How many fibers would suffice for transporting the traffic per connection?

d. How does fewer connections but a larger volume of traffic per connection affect switching?

1.7 References

An interesting and detailed account for the technological, engineering, and operational aspects of the Bell System in the U.S. can be found in the volume edited by Rey. More succinct accounts of telephony and switching can be found in books by Bellamy, Hills, Joel, and McDonald. The book by Inose describes in detail various digital transmission and switching technologies and integrated networks. The excellent books by Bertsekas and Gallager as well as by Schwartz offer a modern look of data networks. Various recent issues of the *IEEE Journal on Selected Areas of Communications* document the latest advances for research in broadband communications.

1. J. Bellamy, *Digital Telephony*, John Wiley and Sons, New York, 1982.

2. D. P. Bertsekas and R. G. Gallager, *Data Networks*, Prentice Hall, Englewood Cliffs, 1987.

3. J. P. Coudreuse, W. D. Sincoskie and J. S. Turner, editors, Issue on Broadband Packet Network, *IEEE J. Selected Areas Commun.*, vo. 6, no. 9, Dec. 1988

4. M. J. Hills, *Telecommunications Switching Principles*, MIT Press, Cambridge and London, 1979.

5. H. Inose, *An Introduction to Digital Integrated Communication Systems*, University of Tokyo Press, Tokyo, 1979.

6. A. E. Joel, editor, *Electronic Switching: Digital Central Office Systems of the World*, IEEE Press, New York, 1982.

7. J. McDonald, editor, *Fundamentals of Digital Switching*, New York: Plenum Press, 1983.

8. H. Ohnsorge, editor, Issue on Broadband Communication Systems, *IEEE J. on Selected Areas Commun.*, vol-4, no. 4, July 1986.

9. R. F. Rey, Editor, *Engineering and operations in the Bell System,* second edition, Bell Telephone Laboratories, 1983.
10. M. Schwartz, *Telecommunication Networks, Protocols, Modeling and Analysis,* Addison Wesley, 1987.
11. P. E. White, J. Y. Hui, M. Decina and R. Yatsuboshi, editors, Issue on Switching Systems for Broadband Networks, *IEEE J. Selected Areas Commun.*, vol-5, no. 8, Oct. 1987.

PART I
Switching Theory

CHAPTER 2

Broadband Integrated Access and Multiplexing

- 2.1 Time Division Multiplexing for Multi-Rate Services
- 2.2 The Synchronous Transfer Mode
- 2.3 The Asynchronous Transfer Mode
- 2.4 Time Division Multiplexing for Bursty Services
- 2.5 Switching Mechanisms by Space or Time Division
- 2.6 Time Multiplexed Space Division Switching
- 2.7 Appendix- What is in a label?

In this chapter, we shall investigate first how services may share a single broadband transmission medium. We move from time multiplexing techniques for single bit-rate services to multiplexing techniques for multi-rate and possibly bursty services. We then describe some mechanisms for time domain and space domain switching, and how these two could be jointly used for switching.

2.1 Time Division Multiplexing for Multi-Rate Services

The problem of multiplexed transmission is stated as follows. We assume that we have m communication terminals $T_1, T_2, \cdots, T_i, \cdots, T_m$ sharing a transmission line. The question then is how we may schedule the sharing of communication bandwidth.

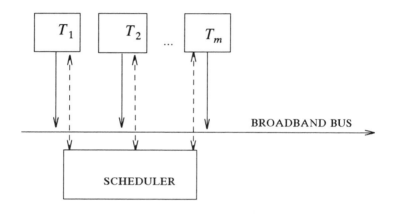

Figure 1. Multiplexing with Scheduling

We assume that the bandwidth is shared by the terminals transmitting at different times. We also assume that a scheduling mechanism is available so that the transmissions are conflict free, namely, that no two terminals attempt to transmit at the same time. We call this scheduled or arbitrated access communication. In the absence of an arbitration mechanism, two communication terminals may transmit at the same time, often resulting in unintelligible transmissions.

Suppose each terminal T_i transmits at a constant bit-rate b_i. The next two sections focus on how these terminals may be multiplexed onto a transmission link. These two sections represent two basic approaches to multiplexing. The first approach assumes a common time reference among the terminals. We call this time reference a frame reference, the function of which will be described later. The communication bandwidth assigned for each terminal is termed a circuit. This mode of multiplexing is commonly known as the Synchronous Transfer Mode (STM).

The second approach assumes no frame reference among the terminals, hence the name Asynchronous Transfer Mode (ATM). This mode allows more flexible sharing of bandwidth by avoiding rigid bandwidth assignments. Bandwidth is seized on demand, and the information transmitted (together with a proper label to be explained later) upon a successful seizure is termed a packet.

We now focus on circuit multiplexing and packet multiplexing in sections 2.2 and 2.3 respectively. The interconnection of circuit multiplexed and packet multiplexed links are called circuit switching and packet switching respectively. This will be the subject of sections 2.5 and 2.6. There is substantial confusion in the popular use of these terms. Besides the vague notion of dedicated bandwidth versus bandwidth on demand, a more precise notion is the presence or absence of a frame reference within the context of time division multiplexing. In the absence of a frame reference, labels for transmitted information become necessary. We shall be guided by this notion in the following discussion.

2.2 The Synchronous Transfer Mode

In order to understand the function of a frame reference, we shall first look at how single rate traffic (namely, all terminals T_i have the same rate b) may be multiplexed.

Suppose the transmission line has a capacity of C bits per second. We may then use the following Time Division Multiplexing (TDM) format for bandwidth sharing:

1. Single-Slot TDM

The transmission capacity can be shared by at most $N = C/b$ transmitting terminals. We mark the infinite time horizon into intervals of some convenient duration d (figure 2). We call each interval a frame. How may we choose a convenient d? Take the telephone network as an example. The network is designed to carry voice signals, which are sampled 8000 times per second. Hence by choosing $d = 1/8000\ sec = 125\ \mu sec$, we may have one speech sample per frame for each voice signal.

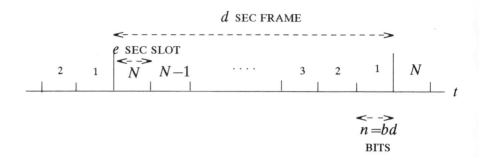

Figure 2. Single-Slot TDM

Each frame is divided into N slots of equal duration $e = d/N$, during which $n = C \times e$ bits can be transmitted. From the point of view of the terminal, a terminal of bit-rate b generates bd bits within the frame duration d. These bits are placed in a chosen slot within a frame. The terminal uses the same slot within consecutive frames. Viewed alternatively, the frame boundaries are time references used in common by the terminals for slot assignments. When a terminal T_i is assigned a slot, we say that it owns a circuit.

2. Multi-Window TDM

However, slot assignment becomes more complicated in a multi-rate environment. Let us assume that there are K types of terminal of bit-rates b_k, $1 \leq k \leq K$. By dividing the channel capacity C into K windows, we have a multi-window TDM format (figure 3).

Each frame of duration d is divided into K windows. Each window has N_k slots each of duration e_k. Consequently,
$$d = \sum_{k=1}^{K} N_k e_k.$$

Within each frame, a type k terminal generates $n_k = b_k d$ bits. These bits are transported in a slot in the k-th window. Therefore, the size of the slot, given by Ce_k, must be at least n_k.

More generally, this multi-window TDM format represents all transmission formats which divide the capacity of a transmission link

2.2 The Synchronous Transfer Mode

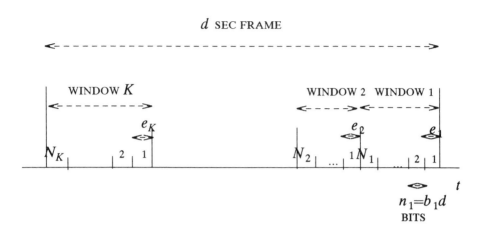

Figure 3. Multi-Window TDM

to accommodate a fixed number of terminal types and a fixed number of services per terminal type. In other words, the network must choose, among the broad range of terminal bit-rates, a set of standard bit-rates $\{b_k\}$, and also the numbers N_k for the windows.

Besides the difficulty of standardizing bit-rates and window sizes in networks with evolving services, there are other problems with this window approach. First, it requires significant effort to format and synchronize the slot and window timings in addition to frame timing, since windows can be of many different sizes. Second, there can be as many windows as there are service or terminal types; otherwise we may have to group several service types of different bit-rates into a window class. Third, the format makes sharing of transmission capacity inflexible among service types.

With respect to the last problem, we may dynamically change the window sizes by moving the window boundaries by changing N_k adaptively, using a technique commonly known as movable boundary TDM. However with more than two windows, adjusting the size of any two windows may necessitate moving many boundaries.

3. Multi-Slot TDM

A more flexible TDM format (figure 4) results from removing the restriction that a terminal may be assigned at most 1 slot within a frame and allowing time slots to be assigned more flexibly. Hence a circuit assigned for a terminal may have many slots.

Let us return to our basic TDM scheme without windows. First choose among the K service types a particular service type, say j, with bit-rate b_j. We use service type j as a reference for deriving the TDM format. Thus there are $N=C/b_j$ slots in a frame, and each slot carries a basic rate b_j. Any service with bit-rate $b_k \leq b_j$ would be assigned exactly 1 slot within each frame. Any service with bit-rate $b_k > b_j$ would be assigned any $\lceil b_k/b_j \rceil$ of available slots within the frame. The slot assignments are the same for consecutive frames.

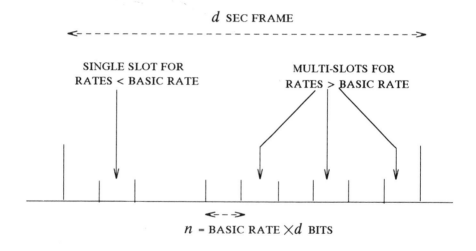

Figure 4. Multi-Slot TDM

Since the slots are of fixed duration, slot timing is easy to derive. However, terminals with lower bit-rate than b_j would not be able to generate enough bits within a frame duration to fill up the assigned slot, resulting in wasted bandwidth. If we choose a small basic bit-rate b_j, then the number of slots $N=C/b_j$ in a frame would be large. Furthermore, terminals with large bit-rate would require a large number of slots. Thus keeping track of the assigned time slot becomes

2.2 The Synchronous Transfer Mode

complicated. Consequently, the basic bit-rate for this multi-slot TDM format has to be chosen judiciously to balance between wasting bandwidth and complicating slot assignment.

Multi-slot TDM, as well as all TDM formats, has the disadvantage that the format depends on the transmission link capacity C, since the number of slots in a frame is given by $N = C/b_j$. However, we may define a basic channel rate C, and view a high capacity link as consisting of multiple channels. The definition of a basic channel rate would make the channel standard independent of technological progress. The basic channel rate should be sufficiently large such that most services would require one channel or less for transport.

2.3 The Asynchronous Transfer Mode

As we have seen, the definition of a frame depends on the bit-rates of the terminals multiplexed on the transmission link. The choice of frame structure is difficult since we have little knowledge of the traffic mix. An alternative approach abandons the concept of a frame reference altogether. Instead of choosing a basic terminal bit-rate as in multi-slot TDM, ATM achieves more flexible bandwidth sharing by allowing the terminals to seize bandwidth when a sufficient number of bits are generated. Without a frame reference, these bits have no implicit ownership, unlike STM for which each slot is assigned an owner. Hence a key feature of ATM is that information from each terminal must be labeled, as we shall explain later.

There are many forms of asynchronous multiplexing. First, we may have fixed length blocks of information from each terminal. These blocks are termed cells in ATM terminology. A cell is a labeled block of transmitted information, and usually has a small information payload (typically from 32 bytes to 128 bytes). We shall also refer to them as short fixed length packets.

1. Cells (Or short fixed length packets)

Each cell or packet has a fixed size of l bits (figure 5). The channel is slotted into fixed intervals of duration l/C, each for transporting a cell. The terminals are asynchronous in the sense that they have no common time reference other than the common slot reference.

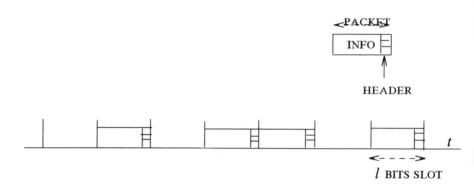

Figure 5. Multiplexing of Fixed Length Packets

A label for each time slot must be provided by the terminal which transmits in that time slot. The label identifies the terminal generating the bits delivered in the time slot. More details concerning the label can be found in the Appendix. A label is included in the header part of a packet. The header may serve other functions, such as classifying the information payload (type and priority), and possibly error check sums for protecting the header from transmission errors.

There are two major factors in determining the proper packet size. First, headers use up part of the communication capacity of the link. This overhead is inversely proportional to the packet size l, consequently favoring long packets. Second, a packetization delay is needed for the terminal to collect the l bits for a packet. The delay between signal generation and reception is given by $\tau_k = l/b_k$, plus the delay taken for the signal to travel in the network. For some applications, excessive delay results in perceivable degradation of the quality of communication. More seriously, delay can be compounded with other network effects such as signal reflections within an analog telephone network to generate undesirable echo effects (see exercises). Consequently, minimizing packetization delay requires choosing short packets. A compromise has to be chosen between these two opposing factors.

For ATM, a terminal tries to seize a time slot whenever it has generated l bits (including the header). However, several terminals

2.3 The Asynchronous Transfer Mode

may try to seize the same time slot. Thus a scheduling mechanism which arbitrates on a slot by slot basis among the contending terminals is required. Priority, which is contained in the header, may be assigned to terminals so that the scheduling mechanism would choose the highest priority terminal contending for a slot.

Instead of a fixed delay comparable with the frame duration for STM, ATM may suffer a random delay which depends on how many packets are contending per slot.

2. Variable Length Packets

Instead of short fixed length packets, it is often convenient (particularly for data communications) to use long (say 128 bytes or more) variable length packets. Besides the label for ownership, the packet header should also contain the information for packet length to mark the end of the packet, as well as a flag to mark the beginning of the packet.

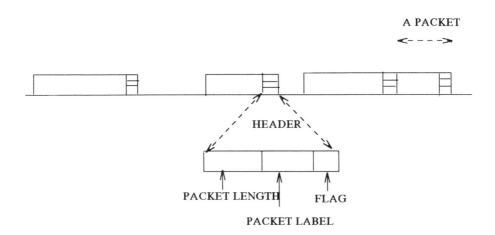

Figure 6. Variable Length Packets

For variable length packets, the variable delay before a packet is successfully transmitted depends on the probability distribution of packet lengths. Thus very long packets may jeopardize the timely delivery of other packets.

For this reason, variable length packets are often fragmented into smaller fixed length packets for transport. In fact, a large variable length packet can be transported via many smaller cells. Hence it is often advantageous to use the term cell instead of the term short fixed length packet for ATM transport, because the transport of a cell does not encompass many functions commonly assumed for the transport of a packet for computer communication. However, we shall use the term packet and not the term cell in later chapters because there is no need for such a distinction, given that we are dealing mainly with the physical aspects of multiplexing and switching.

3. Hybrid TDM (STM + ATM)

Hybrid TDM is a combination of STM and ATM, with the capability of assigning time slots, as well as the flexibility of packet multiplexing. There are many forms of hybrid TDM. For example, we may use the multi-window form of STM, with one of the windows used in an ATM or packet format. However, this form of hybrid TDM is less flexible than a combination of the multi-slot TDM format and the fixed length packet ATM format (figure 7), which we shall describe here.

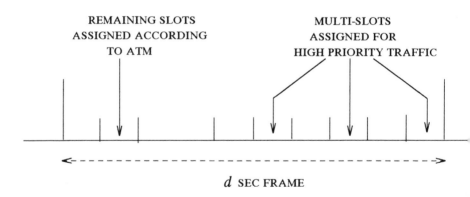

Figure 7. Hybrid TDM = STM + ATM

A frame reference is introduced back into the fixed length packet format. The number of slots in a frame N may not be too large so that we maintain easy tracking of slot assignments. Furthermore, the

2.3 The Asynchronous Transfer Mode

bit-rate $b = C/N$ carried by a slot in a frame need not be tied to a basic terminal bit-rate. If a terminal wants an absolutely fixed delay, one or more time slots may be assigned to the terminal. The terminal might not transmit in all assigned time slots. Such unused time slots could be seized by other terminals on a slot by slot basis. In other words, the terminal which has been assigned a slot has a higher priority for that slot.

Hybrid TDM is introduced to provide the option of circuit switching, which guarantees bandwidth with fixed delay, yet with the flexibility of ATM. The capability to provide circuit switching as needed via priority is important for maintaining compatibility with the existing predominantly circuit switched network.

In summary for this section, labeled (or packet) multiplexing is often the preferred means for realizing integrated access for networks with a broad range of bit-rates. This mode of access avoids the arbitrary choice of a basic terminal rate or frame size among different kinds of terminals. Consequently, new services can be introduced flexibly after the access network is defined and put in place.

2.4 Time Division Multiplexing Techniques for Bursty Services

There is a secondary reason for adopting packet multiplexing; it provides more efficient use of bandwidth when the bit-rate of each terminal is time-varying or bursty. We have assumed so far that the terminals T_i transmit at fixed rate b_i. Now, we shall examine the case that b_i is time varying. In this section we shall see how multi-slot TDM and packet techniques defined in the previous section could be used for variable bit-rate communication. Before we do just that, we shall first look at the traffic processes from communication terminals.

In the previous section, we assumed that a terminal is either not transmitting, or transmitting at a fixed bit-rate. For telephony, a telephone can be off-hook and communicating at a digitized rate of 64 kb/s, or not communicating when on-hook. We have therefore a two state model of the communication terminal (figure 8).

A more careful examination of the off-hook period reveals that there are moments when the speaker is silent. To save communication bandwidth, the telephone may not transmit bits during these silent periods. Thus we may represent the communication process by a 3-

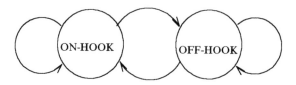

Figure 8. A Two State Model for Telephony

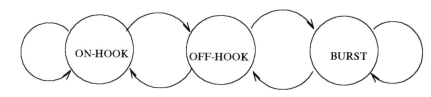

Figure 9. Modeling of Talk Spurts with 3 States

state model (figure 9).

To save transmission cost on expensive links such as transoceanic cables and satellite channels, communication bandwidth is often allocated for each talk spurt. The communication process is initiated when the telephone is taken off-hook. The call initiation is blocked when the communication channel is overly congested. Otherwise, the call request is completed. During the duration of a call, the channel continuously senses whether the call is in a talk spurt state or not. Once a talk spurt is initiated, the channel immediately tries to allocate bandwidth for the talk spurt. For example, the channel may hunt for a vacant time slot in a TDM frame for the talk spurt. The talk spurt is blocked when the channel fails to allocate a circuit for the spurt. This blocking results in clipped speech at the receive end.

Traditionally, this two-layer (call and talk spurt) bandwidth allocation process is known as fast circuit switching. It is viewed as

2.4 Time Division Multiplexing Techniques for Bursty Services

circuit switching because it dedicates bandwidth for individual bursts, and labeled fast because circuits are allocated at the burst level instead of the call level.

We concluded in the previous section that it is more flexible to use packet multiplexing in a broadband network with multi-rate services. Hence with the talk spurt packetized into a multiplicity of fixed length packets, we have a 4-state model for a voice call (figure 10).

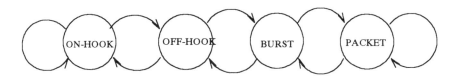

Figure 10. A 4-State Model for Packetized Voice with Silence Removal

When packet switching is employed, it is possible that individual packets within a burst can be lost. Thus on top of the probability of call blocking and burst blocking, we may also have packet blocking. Very often, it is preferable to have one talk spurt lose most of its packets rather than having the same number of lost packets shared by many talk spurts of different conversations, because losing just a few packets of a talk spurt may render the talk spurt unintelligible at the receive end. Similarly, it is often preferable to deny a call request than having many completed calls suffering excessive burst blocking. Thus we have traffic control layered according to the call, burst, and packet layers, with the higher layer controlling the degree of congestion at the lower layer.

Many terminal types other than telephony also exhibit multi-layer bit-rate fluctuations. For still picture applications such as image retrieval from a video database or communication between graphics workstations, the bit-rate process is highly sporadic, with "silent" periods interrupted by high bit-rate communications of pictures. The bursty nature of these terminals bears resemblance to the 3-state model of voice telephony with silence removal. For certain high speed data file transfers and full motion video communications, the bit-rate is constant given a call is completed, hence a 2-state ("on-hook/off-hook") model suffices. Some applications such as compressed full motion video may have a continuously changing bit-rate.

The key feature of an integrated ATM broadband network is the application of statistical averaging to even out the bit-rate fluctuations of the individual terminals. The averaging is facilitated by sharing of large communication bandwidth, and by tolerating statistical delays for satisfying bandwidth requests. This statistical averaging is applied not only to averaging out the random packet arrivals in the case of packet networks, but also to higher layers of random burst and call arrivals.

The modeling and analysis of the bit-rate process will be considered in chapter 7.

2.5 Switching Mechanisms by Time or Space Division

Beyond access, we may want to interconnect transmission links for two reasons. First, the traffic on these links may be multiplexed onto fewer links going to the same location for more efficient utilization of longer distance transmission facilities. We call this process traffic concentration. Second, the traffic on a link may have to be demultiplexed onto transmission links going to different locations. We call this process traffic switching. An interconnection pattern between input and output links is shown in figure 11. Notice that an input may make no connection (e.g. input 4), one connection (e.g. input 1), or more than one connection (e.g. input 2) to the outputs.

A space division switching network can be used to interconnect the incoming links and outgoing links. One configuration of a space division switching network is the crossbar switching matrix shown in figure 12. The crossbar has m inputs and n outputs and mn crosspoint. By making an electrical contact via a crosspoint between a horizontal input bus and a vertical output bus, we make a connection between the associated input and output respectively. The construction of the crosspoint through the use of a logical AND gate is shown in figure 12.

When an input bus makes a contact with more than 1 output bus, the input is multi-casting the same signal to several outputs. We exclude the possibility of more than 1 horizontal bus making a contact with a vertical bus, because the signals from the respective inputs would then be scrambled.

2.5 Switching Mechanisms by Time or Space Division

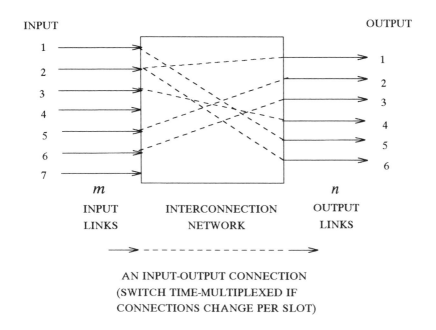

Figure 11. Space Division Switching

Instead of using a space division switching mechanism, time division switching techniques can be applied for interconnecting inputs and outputs. Before we discuss time division switching, we describe a mechanism known as the Time Slot Interchanger (TSI), which is shown in figure 13.

Conceptually, a TSI can be viewed as a buffer which reads from a single input and writes onto a single output. The input is framed into m fixed-length time slots. The numbers contained in each input time slot is the output time slots (none, one, or more than one) for information delivery. The information in each input time slot is read sequentially into consecutive slots (cyclically) of a buffer of m slots. The output is framed into n time slots, and information from the appropriate slot in the buffer is transmitted on the output slot. Hence over the duration of an output frame, the content of the buffer is read out in a random manner according to a read-out sequence as shown in figure 13. This read-out sequence is uniquely determined by the

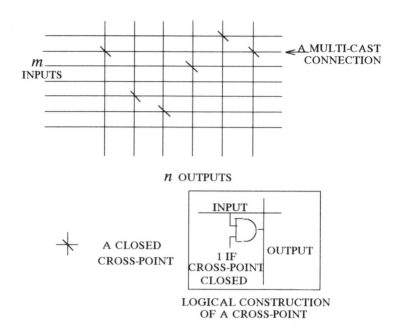

Figure 12. A Crossbar Switching Matrix

connection pattern (indicated by the dotted arrows in the figure.) In this manner, the information in each slot of the input frame is rearranged into the appropriate slot in the output frame, fulfilling the function of time slot interchanging. Also the function of multi-casting can be easily performed by having several output slots reading from the same memory slot.

Time division switching can be performed by a TSI. Since each time slot of a multiplexed link is simply a circuit, the interchanging of information in time slots is equivalent to a switching of circuits. Hence a TSI can also be used to interconnect multiple inputs and outputs (each providing a single circuit), provided that the inputs are first multiplexed onto a single TDM stream, and the time slot interchanged TDM stream from the TSI is then demultiplexed onto the outputs. The space-switched connections via a crossbar switch of figure 12 are realized by the read-in read-out sequence shown in figure 13 for time switched connections via a TSI. Given this functional

2.5 Switching Mechanisms by Time or Space Division

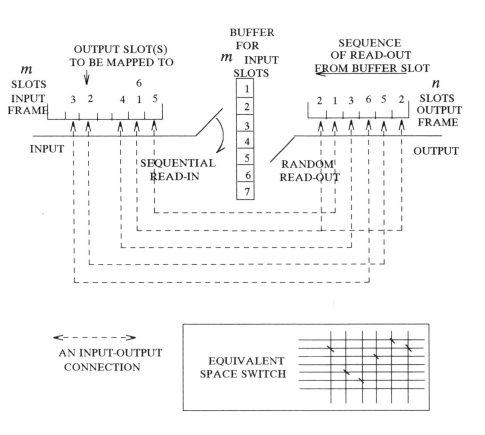

Figure 13. A Time Slot Interchanger

equivalence between time division switching and space division switching, we shall often use the space domain representation to represent either kind of switching.

So far, we have been discussing switching of links which are not time multiplexed. In the next section, we consider the interconnection of time multiplexed (STM or ATM) links. We shall employ both mechanisms of time domain switching and space domain switching described in this section.

2.6 Time Multiplexed Space Division Switching

We now look at how time multiplexed transmission links may be interconnected by a space division switching network such as the crossbar network. We shall first consider the interconnection of STM links before we consider ATM links.

1. Circuit switching for STM links

Let us assume that the frames and time slots are synchronized for the inputs of the crossbar. This can be achieved by using a frame buffer to align the frame timing across the inputs.

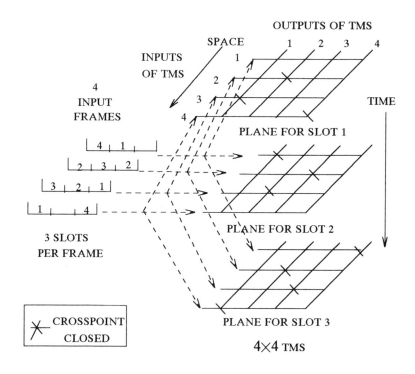

Figure 14. Time Multiplexed Switch (TMS)

For the duration of a slot time, each input may request a connection to an output. The connection requests for each input

2.6 Time Multiplexed Space Division Switching

frame are shown in figure 14. (For example, input 1 requests no connection in the first slot, a connection to output 1 in the second slot, and a connection to output 4 in the third slot.) Suppose the connection requests for the inputs are distinct in each slot. (For example in figure 14, inputs 2, 3, 4 request a connection to outputs 2, 1, 4 respectively in the first slot.) Then we can configure the crosspoint setting of the crossbar switch according to connection requests of the time slot as shown in the plane for slot 1. The crosspoint setting is reconfigured for each time slot. In this sense, we say that the space division switch is time multiplexed, hence the name Time Multiplexed Switch (TMS). We may visualize the time axis vertically, with each switch setting in a slot represented by a horizontal plane as shown in the figure. It can be easily seen that a TMS can be viewed as many space division switches in parallel, by replacing the time axis by a space axis.

Unfortunately, different inputs to a TMS may want to be connected to the same output in the same slot. (For example, input 1 may want to be connected to output 4 in the first time slot. However, this would conflict with the output request of input 4.) We call this phenomenon output conflict. Thus it seems that applying time slot interchanging at each input may alleviate this time slot mismatch problem. Let us consider input 4, which requests a connection to output 4 in the first slot, and output 1 in the third slot. Now suppose input 1 wants to be connected to output 4 also in the first slot. To resolve the conflict for output 4 in the first slot, let us move the 4 in the frame for input 4 to the second slot. Consequently, the output requests for the second slot are 1,3,2,4 for inputs 1,2,3,4 respectively. The move does not cause a conflict in the second slot. Therefore, the conflict for the first slot is resolved!

The application of a TSI for each input to a TMS for resolving output conflicts is shown in figure 15. Since we apply time division switching prior to space division switching, we shall label this as TS switching. Our previous example illustrates how time slot interchanging can remove output conflict. We shall show in the next chapter that a rearrangement can always be made such that the output requests become conflict free for each slot in a frame, provided that the number of requests (per frame) for each output is no more than the number of slots in a frame. (For example in figure 14, there are 2 requests for output 3, which is less than the frame size of 3 slots.) This condition is obviously necessary, otherwise the output would become overloaded.

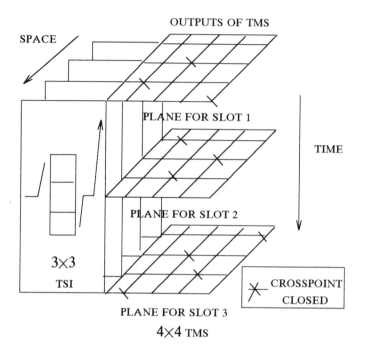

Figure 15. TS Switching for Interconnection TDM links

As mentioned in the previous section, T switching is functionally equivalent to S switching. Also, we can represent a TMS as parallel planes (in time) of S switching. Hence TS switching can be represented by the SS switching shown in figure 16. It is drawn to illustrate the orthogonality of the switching planes between the 2 stages of S switching.

The basic TS switching is sufficient for switching time slots onto the right outputs. However, the slots can appear in any order in the output frame. If we prescribe a specific output slot to carry the information of an input slot, then it is necessary to use a time slot interchanger at each output. In doing so, we can map a connection from a slot of each input to any slot of any output. This 3 stage configuration is termed TST switching, which is equivalent to a 3 stage SSS switching network shown in figure 17.

2.6 Time Multiplexed Space Division Switching

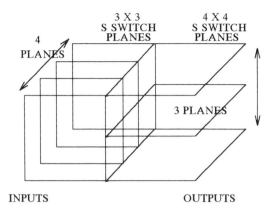

Figure 16. SS Equivalent of TS Switching

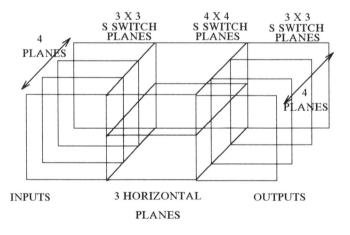

Figure 17. 3 Stage TST or SSS Switching Network

The above connection method can be applied both for single-slot or multi-slot TDM. (For example in figure 14, input 2 requests a 2 slot connection for output 2.)

Before we move on to packet switching, it is noteworthy that the connection of circuits for TST switching is represented by the read-out sequence (figure 13) for each TSI and the crosspoint setting (figure 15) per slot for the TMS. The ownership of slots is implicitly registered by the tables which control these sequencing and setting. We shall focus on these methods of switching control in the next chapter.

2. Packet switching for ATM links

We now consider switching mechanisms for interconnecting ATM links. In particular, we shall focus on the predominant mode of ATM, namely the format using short fixed length packets or cells.

The TMS, used for switching STM links, can also be reconfigured per packet time for connecting ATM links. The same problem of output conflict described for STM arises. As we mentioned before, circuit switching utilizes a TSI at each input of the TMS to remove output conflicts. However, this is impossible for packet switching for two reasons.

First, ATM is inherently an FCFS network without a frame reference. Hence, a packet on an input ATM link would have to wait until the output is available. Consequently, an FCFS waiting room via buffering, instead of memory for time slot interchanging, is necessary at each input.

Second and more important, the output address of packets on an ATM link may not have the periodic structure inherent in an STM network, for which each slot in a frame is dedicated. The periodic structure for STM implies that the crosspoint setting of the TMS is also periodic, hence can be precomputed and stored in a table. However, the crosspoint setting of the TMS is not periodic for ATM and must be recomputed per slot. If the TMS itself is a multi-stage space division switch (say an SSS switch), this computation can be very complicated as we shall show in chapter 3.

By placing FCFS buffer at each input to a TMS, we obtain a packet switch for interconnecting ATM links. We must resolve output conflicts per slot for packet switching, as well as provide efficient means for space division switching of packets if the TMS has a large number of inputs and outputs. These mechanisms will be considered in detail in chapters 5 and 6.

2.6 Time Multiplexed Space Division Switching

2.7 Appendix- What is in a Label?

Let us mention briefly how labels are assigned for ATM. In figure 1, assume that $h < m$ terminals are actively (say when a telephone is taken off-hook) sharing a transmission link. We must use at least $\log_2 h$ bits per label for identifying the active terminals. The label for each terminal is also termed a virtual circuit identifier (VCI). The term virtual distinguishes ATM from the physical and implicit naming of circuits in STM. When a terminal becomes an active user of a link, a unique VCI (with respect to other active terminals using the link) is given to the terminal until it gives up its intention to use the link (say when a telephone is put on-hook). We say that the terminal is assigned a virtual circuit on the link.

The assignment is done per link for all links used by the terminal across a network. Consequently, the VCI assigned on different links can be different, because a terminal shares each link with a different set of terminals. Before the packet is transported to the next link in the network, a table is needed to replace the VCI of a packet on the current link with a new VCI for the next link.

Instead of having different VCI assigned per link, a unique network label may be assigned for each terminal in the network. Hence, no VCI replacement per link is needed. However, many more bits would be required for a network label.

For a variable length packet transported by cells, we have two levels of header. At the cell level, each cell carries the VCI assigned for the terminal. The VCI can be assigned for a terminal and remains the same for many packets. On the other hand, the VCI can be assigned per packet as a packet starts to use a link. Besides the VCI, each cell may carry a cell sequence number marking its sequence in the packet. Very often, the sequence number is represented in modulo arithmetics so that less bits are needed. At the packet level, the first few cells of a packet may be used to transport the information associated with the other header functions of a packet.

Let us examine the process of VCI replacement at a network node such as a packet switch. At the input to the switch, a table replaces the VCI for the current link (an input of the switch) with the VCI for the next link (an output of the switch for which the packet is routed to). Also, the table must direct the packet to the desired output. These tables for ATM serve an analogous function as those for STM, which register the read-out sequence of a TSI and the crosspoint setting for the TMS.

Alternatively, the label may also contain some destination address information beyond that for identification. Many formats of data packets (not the cell format) contain a unique address of the destination terminal in the packet header. The network nodes (a computer or a packet switch) may read the address and provide local routing information for the packet. This mode of packet transport is known as datagram transport which, unlike a virtual circuit, may not require the process of virtual circuit setup and the resulting update of tables in the network node. However, each network node must be capable of interpreting the datagram address and redirecting the datagram to the appropriate output link.

2.8 Exercises

1. Let us study the echo effect for packetized voice.

    a. Consider the transport of 64 kb/s voice over a data network which delivers packets of 1000 bits. What is the packetization delay?

    b. Now suppose we want to call someone on an analog telephone network through the data network, assuming a digital/analog gateway between the data network and the telephone network. Unfortunately, echoes are generated at the far end (say 6000 kilometers away) of the telephone connection. Calculate the echo delay, assuming that the analog signal travels at the speed of light (3×10^8 meters/s).

2. (continuation) Assume there is queueing delay within the packet network, which fluctuates randomly between 2 to 20 packet durations for each packet.

    a. Compute the bounds for the queueing delay for a transmission speed of 10 Mb/s, or 150 Mb/s. How does this delay compare with the delay considered in the previous problem?

    b. How does the randomness of this delay component affect our effort to eliminate the echo?

3. In the multiplexing hierarchy in North America (see reference [2]), some of the important basic channel rates for telephony are DS0 (64kb/s), DS1 or T1 (= 1.544 Mb/s), DS2 (=6.312 Mb/s),

2.8 Exercises

DS3 (= 44.736 Mb/s), and DS4 (=139.264 Mb/s). For optical transmission, the SONET channel rate OTC1 of roughly 50 Mb/s is designed to carry the DS3 information payload, with the difference in channel rate due to extra signaling overhead. Beyond OTC1, channel rates are represented by OTCi, which is formed by byte interleaving of i OTC1 channels.

    a. How many speech (64 kb/s) signals can the DS1, DS2, DS3, DS4 channels carry?

    b. Assume that traffic are multiplexed according to this hierarchy, such that higher rate channels are formed from groups of channels of the next lower rate. Draw a diagram of the hierarchy of multiplexers in order to obtain an DS4=OTC3 channel. Show the number of multiplexers at each level.

    c. We now have multi-layers of frame timing. Discuss schemes of synchronization in this hierarchy. Consider two different approaches of synchronizing or not synchronizing the frames of the multiplexers connected to the multiplexer one level up.

    d. (optional) Discuss the implementation complexity for these two approaches in terms of the memory, synchronization equipment, and the ease of adding/dropping subchannels from a high rate channel. How does using ATM channels help in coping with these issues?

4. Instead of TST switching for which a TSI is placed at each input and output of a TMS, consider STS switching for which two TMS are connected by placing a TSI between each output of the first TMS and the corresponding input of the second TMS. Show that STS switching is equivalent to TST switching. (Hint: show that STS switching is equivalent to SSS switching.)

5. Is it possible to build a TTT switch by using TSI only?

6. Consider the use of 2.4 Gb/s fiber optics link for transporting voice circuits (each requiring 64 kb/s capacity) via ATM cells.

    a. What is the minimum size of the VCI?

    b. What is the minimum size of the VCI if the 2.4 Gb/s link is divided into 16 OTC3 (150 Mb/s) ATM channels?

c. Why can we achieve a reduction in the size of the VCI by having smaller channels? What is the hidden cost?

7. For transporting long variable length packets, we suggested breaking the packets into small fixed length cells so that we may multiplex cells of different packets on an ATM channel. What are the advantage and disadvantage of spacing far apart the cells of a packet?

2.9 References

The framework for network integration and multiplexing methodologies is discussed in a tutorial by Hui. The technological aspects of time division multiplexing and switching are discussed in more detail by Slana. Multi-window and movable-boundary multiplexing schemes were first proposed by Kummerle and Zafiropoulo. One example of multi-slot TDM with standardized basic channel rate is the adopted standard of SONET (Synchronous Optical NETwork) in the U.S., described in the article by Ballart *et. al.* Examples of ATM architectures as well as their hybrid derivatives can be found in the articles by Gonet *et. al.*, Minzer, and Wu *et. al.* New switching architectures exploiting the bursty nature of traffic for improved network efficiency are described by Amtutz and Turner *et. al.* Multi-layer traffic control and analysis are discussed in the articles by Filipiak and Hui.

1. S. R. Amtutz. "Burst switching - an introduction," *IEEE Communications Magazine*, 21(11), pp. 36-42, 1983.

2. R. Ballart, Y. C. Ching, "SONET: now it's the standard optical network," *IEEE Communications Magazine*, March 1989.

3. J. Filipiak, "M-architecture: A structural model of traffic management and control in broadband ISDN", *IEEE Communications Magazine,* vol-27 no. 5, May 1989.

4. P. Gonet, P. Adams, and J. P. Courdreuse, "Asynchronous time-division switching: The way to flexible broadband communication networks," *Proc. IEEE 1986 Int. Zurich Sem. Digital Commun.*, Zurich, Switzerland, March 1986.

5. J. Y. Hui, "Resource allocation for broadband networks," *IEEE J. Selected Areas Commun.*, vol-6, no 9, Dec. 1988.

2.9 References

6. J. Y. Hui, "Network, transport, and switching integration for broadband communications," *IEEE Network Magazine*, vol-3, no 2, pp. 40-51, March 1989.

7. K. Kummerle, "Multiplexer performance for integrated line- and packet-switched Traffic," *ICCC*, Stockholm, 1974, pp. 508-515.

8. S. E. Minzer, "Broadband ISDN and asynchronous transfer mode (ATM)", *IEEE Communications Magazine*, August 1989.

9. M. F. Slana, *Fundamentals of Digital Switching, Chapter 5: Time-Division Networks* (editor: J. C. McDonald), New York: Plenum Press, 1983.

10. J. S. Turner and L. F. Wyatt, "A packet network architecture for integrated services," *Proc. of Globecom'83*, San Diego, 1983.

11. L. T. Wu, S. H. Lee, and T. T. Lee, "Dynamic TDM - A packet approach to broadband networking," *Proc. International Conf. on Commun.*, Seattle, WA, June 1987.

12. P. Zafiropoulo, "Flexible multiplexing for networks supporting line-switched and packet-switched data traffic," *ICCC*, Stockholm, 1974, pp. 517-523.

CHAPTER 3

Point-to-Point Multi-Stage Circuit Switching

- *3.1 Point-to-Point Circuit Switching*
- *3.2 Cost Criteria for Switching*
- *3.3 Multi-Stage Switching Network*
- *3.4 Representing Connections by Paull's Matrix*
- *3.5 Strict-Sense Non-Blocking Clos Networks*
- *3.6 Rearrangeable Networks*
- *3.7 Recursive Construction of Switching Networks*
- *3.8 The Cantor Network*
- *3.9 Control Algorithms*

3.1 Point-to-Point Circuit Switching

In this chapter, we shall deal with the problem of providing point-to-point interconnection between terminals with the same bit-rate such as telephony.

As we stated in the previous chapter, there are two equivalent kinds of switching mechanisms, namely space division switching implemented by crossbar switching and time division switching implemented by time slot interchangers. We have shown that these switching mechanisms are functionally equivalent. Furthermore, if several terminals are first time multiplexed in a TDM format before being switched by a Time Multiplexed Switch (TMS), we need Time Slot Interchangers (TSI) before and after the TMS in order to map an input time slot to an output time slot. The resulting network is a TST switching network, which is functionally equivalent to a three stage SSS space division switching network.

Given this equivalence, we shall use only space division switches in our subsequent treatment of circuit switching. Furthermore, we may represent a space division switch as a node, with inputs represented as incoming edges, and outputs represented as outgoing edges as shown in figure 1.

A space division switching network can be represented by a graph $G=(V,E)$, where V is the set of switching nodes and E the set of edges in the graph shown in figure 2. An edge $e \in E$ is an ordered pair (u,v) for some $u, v \in V$. It is possible that there is more than 1 edge between u and v. We shall count them as distinct edges. Some nodes in V are rather special: let T in V be the set of transmitting terminals which have only outgoing edges, and R in V be the set of receiving terminals which have only incoming edges. Strictly speaking, the set T and R may not be considered as part of the switching network. For the case of duplex connections for which the terminals can both receive and transmit, we may view each edge in E as a bi-directional communication link.

A connection requirement is specified for each $t \in T$ by the subset R_t of R to which t must be connected. The subsets R_t are disjoint for different t. A path is defined as a sequence of connected edges $(t,a), (a,b), (b,c), ..., (f,g), (g,r) \in E$ where $t \in T$, $r \in R_t$, and $a,b,c,...,f,g$ are distinct elements of $V - T - R$, the set of internal nodes. Furthermore, paths originating from different t may not use the same edge. Paths originating from the same t may use the same edges. The set of paths from a t generally forms a tree in case of multi-casting. These connections are illustrated in figure 2. For a given G, the general problem of circuit switching is defined as the process of finding and setting up paths to satisfy a connection requirement.

Point-to-point circuit switching is a special case of multi-cast switching in which R_t contains at most 1 element for each t.

3.1 Point-to-Point Circuit Switching

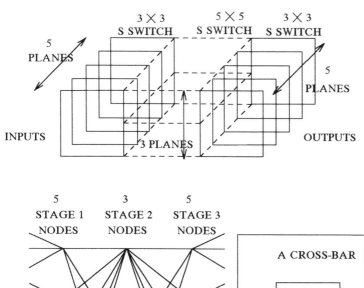

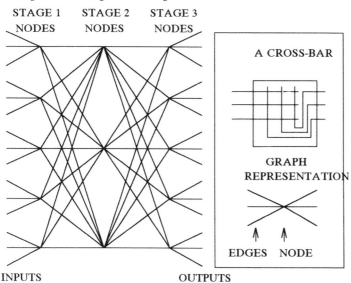

Figure 1. Graph Representation of TST or Equivalent SSS Network

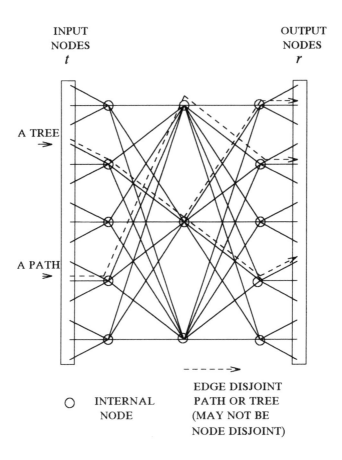

Figure 2. Graph Representation of Connections

3.2 Cost Criteria for Switching

Switching network design involves the choice of the graph G for reducing the overall cost of implementing the switching system. This overall cost consists of a number of components which are highly technology dependent. We shall discuss some general cost considerations for switching network design.

3.2 Cost Criteria for Switching

1. Crosspoint Complexity

Traditionally, most studies on switching networks focus on the number of crosspoints required to implement the switching network. Minimization of crosspoint count is necessary when crosspoint technology is expensive, such as for electro-mechanical or optical crosspoints.

With the economy provided by Very Large Scale Integration (VLSI) technology, a great deal of crosspoint complexity can be implemented on a single chip. In this case, the designer may choose to minimize the chip count instead of crosspoint count. A very important design criterion is the power consumption per chip. This criterion determines how densely the chips can be packed and interconnected.

In many cases, minimizing crosspoint count may not be so important compared with the other cost components. Nevertheless, minimizing crosspoint count is mathematically elegant, and often bears a direct relationship with minimizing power consumption and other cost criteria.

2. Interconnect, fan-out, and logical depth complexity

Given a graph G, we may have to partition the graph into subgraphs each with crosspoint complexity implementable on a VLSI chip. These chips are not limited only in terms of gate counts, but more often by the number of pins per chip, and the fan-in and fan-out capability of each pin. Furthermore, the length of the interconnections between chips is crucial. If the length is too long, the power required to drive the interconnection would be high, or else we may have to lower the speed of the interconnection.

No matter what technology is employed, there is always a limitation on the fan-out of a signal without extensive amplification or regeneration. For example, many local area networks are severely limited by the number of terminals which may be connected to a shared transmission bus before transmission interference or required transmission power becomes excessive.

Logical depth is defined as the number of nodes a signal traverses in the graph. Large logical depth often causes excessive delay and signal deterioration, hence should be avoided if possible.

3. Network control complexity

Given a graph G, we need control algorithms to find and set up paths in G to satisfy the connection requirement between the terminals. Control complexity is defined by the hardware (computation and memory) requirement as well as the run time of the algorithm.

For general G, it is known that the computation required for determining a set of paths for satisfying any connection requirement is generally intractable, in the sense that the computation complexity is potentially exponential in the number of terminals. However, there are G with regular structure for which this control complexity is substantially reduced. There are also classes of G for which the computation can be distributed over a large number of processors. Some switching networks also allow the computation to be localized within the neighborhood of the switching node. Such networks allow rapid updates of paths for services with volatile connection requirement.

In general, the amount of computation depends also on the degree and kinds of blocking we may tolerate.

4. Degree and kinds of blocking tolerated

Generally speaking, blocking is defined as failure to satisfy a connection requirement. Such a failure can be attributed to different causes.

The earlier kinds of switching networks employed for telephony are blocking in the sense that not all connection requirements can be satisfied, simply because non-conflicting paths do not exist for some connection requirements. Thus a telephone call can be blocked even though the 2 terminals involved are free. This kind of blocking depends strongly on the combinatorial properties of the graph G.

The broadest class of non-blocking networks is that of the so-called rearrangeable networks, for which a set of non-conflicting paths always exists for any connection requirement. It is termed rearrangeable because the path of a connected pair of terminals may have to be changed or rearranged when some other pair of terminals initiates a connection. Rearranging paths for connected pairs involves extra control complexity, and may cause disruption of a connection.

A non-blocking network is termed strict-sense non-blocking if a path can be set up between an idle transmitter and an idle receiver (can be more than one receiver for the case of multi-casting) without

3.2 Cost Criteria for Switching

disturbing the paths already setup. There is always a free path for any connection initiated.

In between these 2 classes of non-blocking networks, there are networks called wide sense non-blocking networks for which new connections can be made without disturbing existing connections, provided we follow some rules for choosing a path for the new connection. In other words, these rules prevent the network from getting into a state for which new connections cannot be made. Very little is known so far about wide sense non-blocking networks.

In practice, we may tolerate a small probability of blocking. In return, we reduce the crosspoint complexity and control complexity. Many earlier switching systems are combinatorially incapable of realizing all connection requirements. (For example, the number 5 crossbar network described in exercise 1 of chapter 4.) Most recent switching systems are either rearrangeably non-blocking or strict-sense non-blocking. However, blocking still arises in practice because the control algorithms very often do not perform an exhaustive search for free paths or rearrangements of paths.

5. Network management complexity

The previous complexities determine the kind of graph and control algorithm put in place for a switching system. Network management involves the adaptation and maintenance of the switching network after the switching system is put in place, to cope with failure events and growth in connectivity demand in the network. Also, network management deals with the changes of traffic patterns from day to day, as well as momentary overloads. The choice of the switching network topology and control algorithm can have a significant impact on network maintenance cost and the incremental cost for network growth.

A modern switching system must be capable of diagnosing the occurrence of hardware failure in the switching network, the control system, and the access and trunk networks. Once a failure is detected, the switching system automatically reroutes traffic through redundant built-in hardware, or reroutes via other switching facilities. Such diagnostic and failure maintenance measures constitute a significant part of the software programs of a switching system.

Ideally, we would like the capacity of a switching system to be capable of growing gradually as traffic demand increases. In order for switching cost to grow linearly with respect to total traffic, switching functions such as control, maintenance, call processing and the interconnection network should be as modular as possible. New modules could be plugged into the system as demand increases.

3.3 Multi-Stage Switching Networks

In general, deciding whether a network G is non-blocking and finding non-conflicting paths in G is difficult. However, efficient algorithms are known for certain multi-stage switching networks with very regular structures.

We may use a single $N \times N$ crossbar interconnecting N transmitting terminals to N receiving terminals. However, such an array would require N^2 crosspoints. More serious than this large crosspoint count is that each input and output is of linear (in N) length and may be affected by N crosspoints. Given the limited driving power of electronics and optics, the propagating signal may suffer deterioration in the absence of signal regeneration.

A multi-stage network consists of K stages of switching nodes. Stage k ($1 \leq k \leq K$) has r_k switches. Each switch in stage k is named by a number j, $1 \leq j \leq r_k$. We represent that switch by $S(j,k)$. Also, such a switch has m_k input edges and n_k output edges. We represent the input i of $S(j,k)$ by $e(i,j,k)$, and the output i of $S(j,k)$ by $o(i,j,k)$. The interconnection between stages is specified by the way the switches between consecutive stages are connected. In other words, we specify which output of a switch is the input to another switch in the next stage by relationships $o(i,j,k) = e(i',j',k+1)$.

In this chapter, we shall look at a very special class of interconnections: each switch in each stage is connected to each switch in the next stage by 1 edge. Thus, the number of outgoing edges $n_k = r_{k+1}$, the number of switches in the next stage. Likewise, the number of incoming edges $m_k = r_{k-1}$, the number of switches in the previous stage.

The goal of this chapter is to demonstrate efficient point-to-point circuit assignment algorithms for this special class of multi-stage networks. However, we should be very careful about what we mean by efficient. Towards the end of the chapter, we shall come to the conclusion that even though these algorithms are efficient for circuit switching, they are not fast enough for packet switching.

3.4 Representing Connections by Paull's Matrix

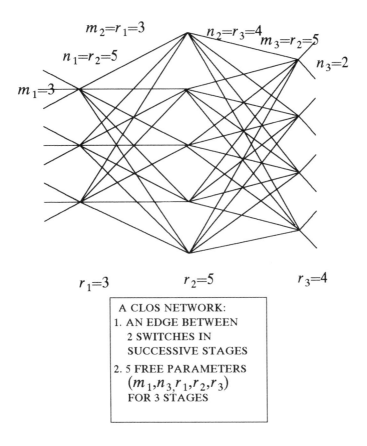

Figure 3. Size of Clos Networks

In this section, we shall only be dealing with three stage networks, with each switch in a stage connected to each switch in the next stage by one edge. Recall that a three stage network is defined by five parameters: r_1, r_2, r_3 which are the number of switches in each stage, and m_1 the number of inputs to a first stage switch, as well as n_3, the number of outputs to a third stage switch. The other parameters $n_1=r_2$, $m_2=r_1$, $n_2=r_3$, and $m_3=r_2$ are then uniquely determined since switches in consecutive stages are connected by exactly one edge. This network, first studied by Clos and therefore named a Clos network, is shown in figure 3.

We shall use a matrix notation devised by Paull for representing paths in the network. Consider any switch a in stage 1 and any switch b in stage 3. An input to a may be connected to an output of b via a middle switch f. Other inputs to a may be connected to other outputs of b via the middle switches g, h, etc., as shown in figure 4.

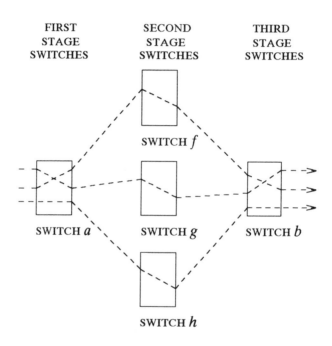

Figure 4. Path Connections in 3-Stage Network

Paull's connection matrix (figure 5) represents these paths by entering the middle switches f, g, h etc. into the (a,b) entry of an $r_1 \times r_3$ matrix. It should be noted that each entry of the matrix may have none, one, or more than 1 middle switches.

Consider first the problem of point-to-point switching. The conditions (figure 6) for a legitimate connection matrix are given by:

3.4 Representing Connections by Paull's Matrix

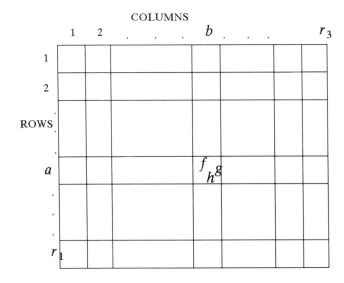

Figure 5. Paull's Matrix for Representation of Connections

1. Each row can have at most m_1 symbols, since there can be at most as many paths through a first stage switch as there are inputs to that switch.

2. Each column can have at most n_3 symbols, since there can be at most as many paths through a third stage switch as there are outputs to that switch.

3. The symbols in each row must be distinct, since we have only 1 edge from the first stage switch to any second stage switch, and we do not allow multi-casting to different third stage switches from a second stage switch. Therefore, there can be at most r_2 symbols.

4. The symbols in each column must be distinct, since we have only 1 edge between a second stage switch and a third stage switch and an edge cannot carry signals from two different inputs. Therefore, there can be at most r_2 symbols.

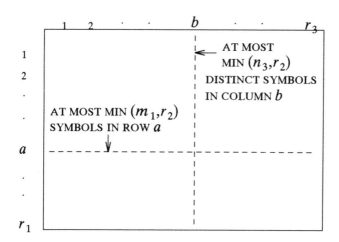

Figure 6. Conditions on Paull's Matrix for Point-to-Point Connection

When we allow multi-casting, condition 1 and 3 may not be valid because a path from a first stage switch may fan-out into a multi-cast tree at the second stage switch. However, conditions 2 and 4 remain valid.

3.5 Strict-Sense Non-Blocking Clos Networks

Let T' be any subset of set T of transmitting terminals and R' be any subset of the set R of receiving terminals. Each element of T' is connected by a legitimate multi-cast tree to a non-empty and disjoint subset of R'. Each element of R' is connected to one element in T'.

A network is strict-sense non-blocking if any $t \in T-T'$ can establish a legitimate multi-cast tree to any subset of $R-R'$ without changing the previously established paths, for all T' and R' as well as for all connection patterns between T' and R'. Strict-sense non-blocking networks have the desirable property that existing connections need not be disturbed in the process of setting up a new connection. More generally, a rearrangeable network satisfies the

3.5 Strict-Sense Non-Blocking Clos Networks

same condition, except that changing the previously established paths is allowed.

The condition for a Clos network to be strict-sense non-blocking for point-to-point connections is given by

Clos Theorem:

A Clos network is strict-sense non-blocking if and only if the number of second stage switches $r_2 \geq m_1 + n_3 - 1$.

In particular, a symmetric network with $m_1 = n_3 = n$ is strict-sense non-blocking if and only if $r_2 \geq 2n - 1$.

Proof:

Suppose we want to establish a connection from a vacant input of a first stage switch a and a vacant output of a third stage switch b. We do so by putting a symbol in the (a,b) entry in Paull's connection matrix. Now there can be at most $m_1 - 1$ distinct symbols in row a, because there are only m_1 inputs to switch a, less one input which wants to make a new connection. For a similar reason, there can be at most $n_3 - 1$ distinct symbols in column b. Consider the worst case when these $(m_1 - 1) + (n_3 - 1)$ symbols are all distinct. If we have 1 more second stage switch, in other words a total of $m_1 + n_3 - 1$ second stage switches, then we can enter a symbol into (a,b) while keeping all symbols in row a distinct, and all symbols in column b distinct as well.

This condition on r_2 is also necessary by considering the following sequence of connections. Suppose we first make connections from a single first stage switch a to all third stage switches. In the process, we enter symbols into row a of the matrix. Then we make connections from a single third stage switch b to all first stage switches except a. In the process, we enter symbols into column b. Obviously, we need a total of $m_1 + n_3 - 1$ distinct symbols. □

The preceding proof also gives the procedure for making connections shown in figure 7. We can keep track of the symbols used by row a using an occupancy vector $\underline{u}_a$ with r_2 entries. We put a 1 in the position a of the vector if the symbol a has been used in the row; otherwise, we put a 0. Likewise, we keep track of the symbols used by column b by the vector $\underline{v}_b$. To make a connection between switches a and b, we look for a position in $\underline{u}_a$ and $\underline{v}_b$ both with an entry 0.

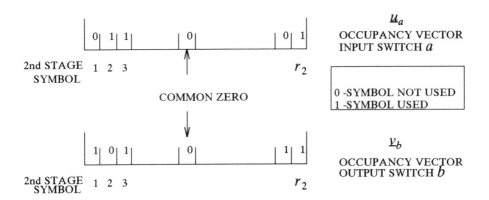

Figure 7. Slot Matching for Circuit Switching

The amount of computation required is proportional to r_2.

3.6 Rearrangeable Networks

In this section, we consider the number of second stage switches needed to provide point-to-point connections, given that we may rearrange connections that are already established. We shall give the necessary and sufficient condition for a Clos network to be rearrangeably non-blocking. The proof also gives a specific algorithm for rearranging connections. Therefore, we can also estimate the complexity of the rearrangement process.

The Slepian-Duguid Theorem:

A three stage Clos network is rearrangeable if and only if $r_2 \geq \max(m_1, n_3)$.

In particular, a symmetric network with $m_1 = n_3 = n$ is rearrangeably non-blocking if and only if $r_2 \geq n$.

Proof:

3.6 Rearrangeable Networks

Suppose we want to establish a connection between switches a and b. We shall first prove that if $r_2 \geq \max(m_1, n_3)$ we must either have

1. A symbol which is not found in both row a and column b; or
2. There exists a symbol c in row a which is not found in column b, and a symbol d in column b which is not found in row a. (See figure 8.)

For if case 1 is not true, the r_2 symbols must be found either in the row or the column or both. Using the assumed condition on r_2, we have $r_2 > m_1 - 1$. Now there are at most $m_1 - 1$ symbols in row a. Hence there must be a d in column b not found in a. Using a symmetrical argument, there must be a c in row a not found in b.

If case 1 is true, then we can place the unfound symbol in (a, b), thus completing the connection without any rearrangements. Otherwise, we look at the row where d appears in column b to see if the symbol c appears in that row as shown in figure 8.

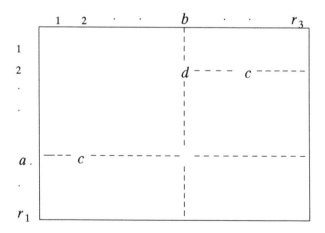

Figure 8. Row-Column Search Process

If such a c is found, we check that column to see if a d appears, as shown in figure 9. We continue the process alternately until we cannot find a c or d. The rearrangement is facilitated by putting a d

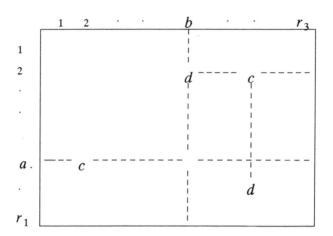

Figure 9. Growing Alternating Chain of Symbols

in (a,b), and replacing all c with d as well as all d with c in the chain as shown in figure 10. Obviously, the four conditions for a legitimate connection matrix is satisfied with such rearrangements.

Since there are m_1 inputs to a first stage switch and n_3 outputs to a third stage switch, it is also necessary to have at least $\max(m_1,n_3)$ second stage switches in order that the first or third stage switches may be fully utilized. ☐

Since there can be at most a total of r_1+r_3-2 rows and columns in the matrix which can be visited by the chain, there are at most that many circuit rearrangements. We subtract 2 because the row a which contains a symbol c needs no rearrangement, and the last symbol in the chain visits both a column and a row.

Paull reduced the number of rearrangements, as stated in the following theorem:

Paull's Theorem:

The number of circuits that need to be rearranged is at most $\min(r_1,r_3)-1$.

3.6 Rearrangeable Networks

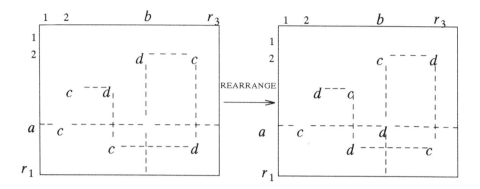

Figure 10. The Process of Rearrangement

Proof:

Without loss of generality, let us assume that $r_3 \leq r_1$, namely that there are fewer columns than rows in the Paull's matrix. First, we look for a c in the row for the d that appears in column b as shown in figure 8. Next, we start the chain from c in row a, instead of extending the chain from d in column b. We now extend the two chains alternately as shown in figure 11, so that at each step, the chains have lengths differing by at most one. When either one of the chain cannot be grown further, we choose that chain for rearrangement.

In each step for which the length of both chains are increased by one, we visit one new column. Hence there can be at most r_3-2 steps in extending the two chains. (We subtract two because the initial c and d occupy two columns.) Hence we need at most r_3-1 rearrangements. (The initial c or d also has to be rearranged.)

The same argument applies when we have $r_3 \geq r_1$. Hence we need at most $\min(r_1, r_3) - 1$ rearrangements   □

Notice that in the worst case, the use of rearrangeable networks instead of strict-sense non-blocking networks increases the circuit setup complexity by roughly a factor of r_1 or r_3. However, the average number of circuit rearrangements may be substantially less if the

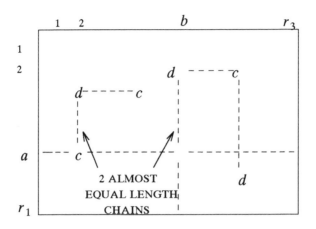

Figure 11. Extending 2 Chains Simultaneously

switching network is not heavily loaded.

3.7 Recursive Construction of Switching Networks.

We have so far factored a switching network into three stages of smaller switching networks, and have given conditions on the number of second stage switches for strict-sense or rearrangeably non-blocking networks. We can carry this three stage factorization further for these smaller switching networks, for the purpose of reducing their crosspoint complexity.

Suppose we want to construct an $N \times N$ switching network. Let $N = p \times q$. Strict-sense and rearrangeably non-blocking networks can be constructed in the manner shown in figure 12.

The crosspoint count for the rearrangeable construction is

$$p^2q + q^2p + p^2q = 2p^2q + q^2p \qquad (3.7.1)$$

The crosspoint count for the strictly non-blocking construction is

$$p(2p-1)q + q^2(2p-1) + p(2p-1)q = 2p(2p-1)q + q^2(2p-1) \qquad (3.7.2)$$

3.7 Recursive Construction of Switching Networks

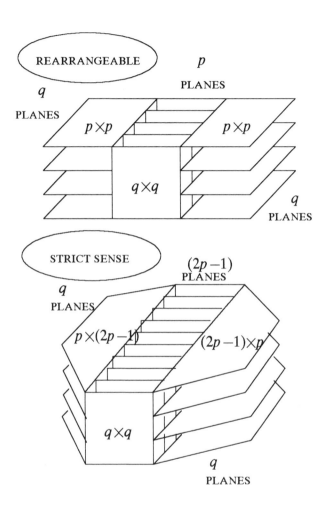

Figure 12. 3-Dimensional Non-Blocking Networks

Notice that these expressions are not symmetrical in p and q.

A recursively constructed rearrangeably non-blocking Clos network is formed by implementing each $p \times p$ or $q \times q$ switch (figure 12) as a three-stage rearrangeably non-blocking network. This recursive construction can be repeated for the smaller switches.

A recursively constructed strict-sense non-blocking Clos network is formed by implementing each $p \times (2p-1)$ or $q \times q$ switch (figure 12) as a three-stage strict-sense non-blocking network. This recursive construction can also be repeated for the smaller switches.

Suppose N can be factored into p and q in many ways, which can be further factored. Which factor p should be chosen first, and how should the subnetworks be further factored? The doubling due to identical first and third stages suggests that we should start with the smallest factor (necessarily a prime number), and recursively factor $q = N/p$ using the next smallest factor. This strategy seems to work fairly well for rearrangeable networks. For strict-sense non-blocking networks, there is a doubling in the width of the network as shown in figure 12. Thus recursively using the smallest factor first also causes the fastest growth in width, defined as the number of the nodes in the center stage of the network. Therefore, choosing the smallest factor first may not be the best strategy for minimizing the crosspoint count.

For the special case of $N = 2^n$, n being a positive integer, we can recursively construct a rearrangeable network by factoring N into $p = 2$ and $q = N/2$. The resulting network shown in figure 13 is called a Benes network, which has $2\log_2 N - 1$ stages. Each stage consists of $N/2$ switches of size 2×2. Therefore, the number of crosspoints according to this recursive construction is roughly $4N \log_2 N$. As we shall see in chapter 4, this is roughly the minimum number required.

A baseline network is defined as the first half of the Benes network, namely the left part of the network from the inputs of the Benes network to the outputs of the binary switch nodes at stage $\log_2 N$. An inverse baseline network is defined as the second half of the Benes network, namely the right part of the network from the inputs of the binary switch nodes at stage $\log_2 N$ to the outputs of the Benes network. The baseline networks belong to the more general class of banyan networks, which will be discussed in more detail in chapter 5.

Unfortunately, the recursive construction of a strict-sense non-blocking network does not give a network with fewer than $CN \log_2 N$ crosspoints, for all N and a fixed constant C. Instead of starting with the smallest factor, let us factor N into three stages, each consisting of switches of size $\sqrt{N} \times \sqrt{N}$. For convenience, let us assume that $N = 2^n$ and $n = 2^l$ so that in each step of a recursion, we are factoring square switches with the number of inputs and outputs being a power of two.

3.7 Recursive Construction of Switching Networks

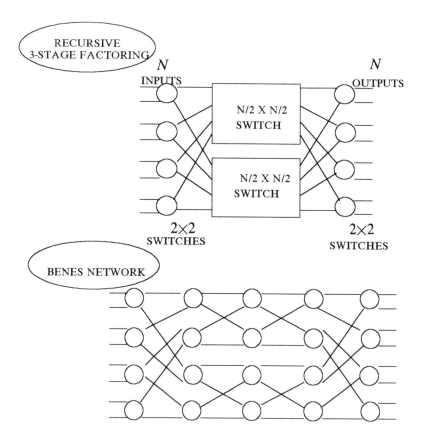

Figure 13. Recursive Factorization for the Benes Network

The condition for strict-sense non-blocking is that there are at least $2 \times 2^{n/2} - 1$ second stage switches. For convenience, we assume there are $2 \times 2^{n/2}$ such switches. The first stage switches are of size $2^{n/2} \times 2^{n/2+1}$, and the third stage switches are of size $2^{n/2+1} \times 2^{n/2}$. Each of these switches can be made from 2 switches each of size $2^{n/2} \times 2^{n/2}$. The second stage switches are of size $2^{n/2} \times 2^{n/2}$. Therefore, the three stages consist of $6 \times 2^{n/2}$ switches of size $2^{n/2} \times 2^{n/2}$. Let $F(2^n)$ be the crosspoint complexity of the $N \times N$ switch. We have then the recursive relationship:

$$F(2^n) = 6 \times 2^{n/2} F(2^{n/2}) \qquad (3.7.3)$$

$$= 6^l 2^{n/2+n/4+\ldots+1} F(2^1)$$

$$< 6^l 2^n F(2)$$

$$= N (\log_2 N)^{2.58} F(2)$$

(Note : $2.58 = \log_2 6$.)

The difference in complexity in the above constructions for rearrangeable and strict-sense non-blocking networks lies in the exponent for the log$N$ term. Are there strict-sense non-blocking networks with this exponent less than 2.58, or even approaching 1? The answer is yes, using entirely different approaches.

3.8 The Cantor Network

The first alternative approach is the Cantor network. Each input and output of the Cantor network are connected to the corresponding input and output of m Benes network (figure 14) through demultiplexers and multiplexers. In other words, the i-th input of the Cantor network is connected to the j-th input of the j-th Benes network using the j-th output of a $1 \times m$ demultiplexer. Similarly, the i-th output of the j-th Benes network is connected to the i-th output of the Cantor network through the j-th input of a $m \times 1$ multiplexer.

We now show that $m = \log_2 N$ Benes planes (as shown in figure 14) is sufficient to make the Cantor network strict-sense non-blocking. Since the Benes network has crosspoint count $4N\log_2 N$ the Cantor network gives a strict-sense non-blocking network of complexity of roughly $4N(\log_2 N)^2$, if we ignore the crosspoint count for the multiplexers and demultiplexers.

Proof:

Each Benes network has $2\log_2 N - 1$ stages. Consider the number of switching nodes for all m Benes networks reachable (without rearrangement) by an input of the Cantor network. We define $A(k)$ as the number of nodes of the Benes network reached by the input at stage k, $1 \leq k \leq \log_2 N$. Obviously, each input can reach one node in each of the m Benes planes, hence

$$A(1) = m \qquad (3.8.1)$$

3.8 The Cantor Network

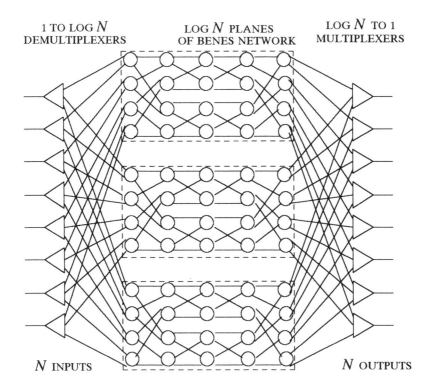

Figure 14. The Cantor Network

These $A(1)$ nodes are also reachable by another input (check figure 14). This other input uses up one of the $2A(1)$ outgoing links from these $A(1)$ nodes. Hence

$$A(2) = 2A(1)-1 \qquad (3.8.2)$$

Similarly, these $A(2)$ nodes can reach $2A(2)$ stage 3 nodes. Two of these $A(2)$ nodes are reached by two other inputs of the Cantor network not considered before. Hence

$$A(3) = 2A(2)-2 \qquad (3.8.3)$$

An input can reach $2A(k-1)$ nodes at stage k, less 2^{k-2} nodes reached by 2^{k-2} inputs of the Cantor network not accounted for in the previous calculations. Hence

$$A(k) = 2A(k-1) - 2^{k-2} \qquad (3.8.4)$$

$$= 2^2 A(k-2) - 2 \times 2^{k-2}$$

$$= 2^{k-1} A(1) - (k-1)2^{k-2}$$

Hence the number of reachable nodes at stage $k = \log N$ is

$$A(\log N) = 2^{\log N - 1} m - (\log N - 1) 2^{\log N - 2} \qquad (3.8.5)$$

$$= \frac{1}{2} Nm - \frac{1}{4}(\log N - 1) N$$

Since the Cantor network is symmetrical at the middle, the same number of the center stage nodes are reachable by an output of the Cantor network. There are a total of $Nm/2$ nodes in the center stage of the m Benes networks. If the sum of center stage nodes reachable by an input and an output exceeds $Nm/2$, there must exist a node reachable from both the input and the output considered. Hence strict-sense non-blocking is achieved if

$$2 \times \left[\frac{1}{2} Nm - \frac{1}{4}(\log N - 1) N \right] > \frac{Nm}{2} \qquad (3.8.6)$$

which reduces to the condition

$$m > \log N - 1 \qquad (3.8.7)$$

Therefore, it suffices to have $m = \log N$ Benes planes so that the Cantor network is strict-sense non-blocking. □

Finally, it is worth mentioning that there is an existence proof for strict-sense non-blocking networks with complexity less than $CN \log N$ for some constant C. We do not describe these networks for several reasons. First, it is not clear that such networks have a regular and easily implementable structure. Second, the constant for C is so large that in comparison, the strict-sense non-blocking Clos network has less complexity for all practical purposes. Third, very little is known about the control algorithm for these networks.

Control complexity is often of equal importance to the crosspoint complexity of a network. We shall look at control algorithms for these multi-stage networks in the next section.

3.9 Control Algorithms

Control algorithms for strict-sense and rearrangeably non-blocking networks are implicitly stated in the previous proofs (see sections 3.5 and 3.6) for strict-sense and rearrangeable conditions. In this section, we shall examine control algorithms for networks which are formed recursively by three stage factorization.

Control algorithms can be applied recursively. We first determine the input-output connections on each switch of the three stages before we determine the connections on further factorization of a switch in a stage. For strict-sense non-blocking networks, this recursive application works very well when we make connections one at a time. For rearrangeable networks, the connection pattern across the whole switching network can be dramatically changed, even though we want to add just one connection. Consequently, adding a connection in a Benes network can be almost as complicated as reconnecting all input-output pairs.

Let us reexamine the control algorithm for the following three stage rearrangeable network:

1. There are $N = 2^n$ inputs and outputs, with the network factored recursively to form the Benes network.
2. We shall start with a totally disconnected network, and establish the connection pattern requested.

One method of establishing the connections is via the looping algorithm. During the first factorization of the $N \times N$ network, each 2×2 input switch may be connected to a 2×2 output switch either by the upper or lower $N/2 \times N/2$ networks (figure 15). In other words, we have two kinds of symbols, namely U and L for entry in the Paull's matrix.

The Looping Algorithm:
1. (Initialization)
   Start with input switch 1. Denote this 2×2 switch node by S.

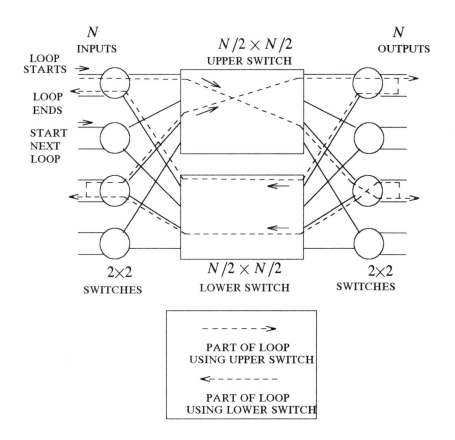

Figure 15. The Looping Algorithm

2. (Loop forward)
   Connect an unconnected input of S to the desired output by the upper switch U. If no connection is needed, go to 4.

3. (Loop backward)
   Connect the adjacent (i.e. on the same 2×2 switch) output of the output just visited to the desired input by the lower switch L. If no connection is needed, go to 4. Otherwise, the newly visited input switch becomes S. Go to 2.

3.9 Control Algorithms

4. (Start new loop)
   Choose another input switch which has not been visited before as S. Go to 2. If all connections for the $N \times N$ switch are made, the algorithm terminates at level n.

This algorithm is illustrated in figure 15. Notice that each loop in this algorithm corresponds to a loop in Paull's matrix.

The same algorithm can be applied recursively to establish the connections for the upper switch U and lower switch L. Hence the algorithm can be used to set up connections for a Benes network. Let us consider the use of this algorithm for packet switching via a Benes network. At the beginning of each time slot, we have packets at the inputs of the switching network. Let us assume that these packets must be delivered to distinct outputs. We want to investigate the speed at which the paths can be computed. In particular, we discuss briefly the use of parallel processing to speed up the looping algorithm.

Suppose each input and output binary switch simultaneously try to find out the other binary switches on the same loop. It can be shown that this can be accomplished in time $\log N$. Given that we know the switches on the loop, we may use the upper switch U for one of the connections in the loop. Once we connect a link of a loop by U, the switch (U or L) used by the other links of the loop is immediately determined. Thus the switches on a loop need to propagate this common agreement of the initial choice of the upper or lower switch for a particular link. The propagation of this agreement also requires $\log N$ time.

To perform the above computations, we use a single processor for each input and output 2×2 switch. These processors are fully connected to each other. Using this configuration, it is possible to distribute the computation of loops and propagate the link assignments in order $\log_2 N$ time. This process is repeated recursively for the $N/2 \times N/2$ subnetworks, using the same processors. The total run time of the algorithm is roughly proportional to $(\log_2 N)^2$.

This parallel implementation is seldom used for two reasons. The time complexity is not sufficiently small, and the necessity of full connection between N processors defeats the purpose of the algorithm of trying to to provide interconnection between N inputs and N outputs without full N^2 interconnection in the first place.

We describe this parallel looping algorithm here for the purpose of illustrating that even though this circuit setup algorithm (as well the previously described algorithms) may be efficient enough for circuit switching, it is not fast enough for packet switching. In circuit switching, we only have to compute a connection per call, while in packet switching, we may have to recompute the interconnection for all N input-output pairs within the duration of a packet. Given the high complexity of the parallelized looping algorithm, we have to examine alternative switching network architectures for packet switching, which is the focus of later chapters.

3.10 Exercises

1. Compute the crosspoint complexity, logical depth (the number of logical gates in a path), and fan-out (the number of logical gates driven by the input or by any gate in the network) for the following networks.
   a. The full $N \times N$ crosspoint switch.
   b. The three stage rearrangeable Clos network constructed using $\sqrt{N} \times \sqrt{N}$ switches.
   c. The Benes network.

2. Consider the crosspoint complexity of three stage Clos networks.
   a. Show that the strict-sense network has roughly twice the complexity of the rearrangeable network.
   b. For the rearrangeable network, show that the optimal choice of p (figure 12) for minimizing crosspoint count is $\sqrt{N/2}$, which gives a crosspoint complexity of $2\sqrt{2}N^{3/2}$.
   c. For the strict-sense network, show that the minimum crosspoint count is roughly given by $4\sqrt{2}N^{3/2}$.

3. Consider the following scale up by a factor of l for a three stage factoring of a rearrangeable network via $N = p \times q$ as shown in figure 12. Suppose now we replace each edge by l edges, each $p \times p$ switch by an $lp \times lp$ switch, and each $q \times q$ switch by an $lq \times lq$ switch. Show that the resulting $lN \times lN$ network is rearrangeable. Furthermore, show that the above scale up has unnecessary crosspoints for constructing an $lN \times lN$ rearrangeable network. (Hint: Consider an equivalent $lN \times lN$

3.10 Exercises

Clos network.)

4. A rearrangeable SSS network with a TSI at each link in the network (including the input links, the output links, and the internal links) is termed a TSTSTST network.

   a. Show that the resulting network is rearrangeable, for which any input time slot can be connected to any output time slot.

   b. Can a multi-slot connection from an input to an output in the SSS network use the same spatial path if a TSI is placed at each link in this TSTSTST network?

5. Let us compare the complexity of the Cantor network with the complexity of the strict-sense non-blocking Clos network.

   a. Accounting for the crosspoint count for multiplexers and demultiplexers, show that the crosspoint count for the Cantor network is $4N(\log_2 N)^2$.

   b. Compute $N=2^n$ for which an $N \times N$ Cantor network has a smaller crosspoint count than that of the optimal three stage strict-sense non-blocking network considered in problem 2.

6. Consider the recursive construction of an $N \times N$ rearrangeably non-blocking Clos network using $p \times p$ crossbars only.

   a. Compute the number of crosspoints as a function of N and p.

   b. For large N, show that $p=3$ minimizes the crosspoint count.

7. Explain quantitatively why there is a bigger chance of finding a common zero (figure 7) when the number of slots per frame is large, given a constant fraction of the slots are occupied for the input and output frames. (For simplicity, you may assume each slot is occupied with a fixed probability p.) Using this result, comment on the use of rearrangeable networks without allowing circuit rearrangements.

8. For the 8×8 Benes network, use the looping algorithm to find the paths for the following permutation:

input	1	2	3	4	5	6	7	8
output	3	6	2	1	8	4	5	7

9. Suppose you are given a permutation of $(1,2,3,\cdots,N)$, such as the output permutation shown in the previous problem. Let us consider automating the looping algorithm.

   a. Write a computer program which returns an N-vector for which the i-th entry is either U or L, denoting whether the i-th input should be connected to the corresponding output in the permutation via the Upper switch or the Lower switch.

   b. Estimate the time complexity of your program.

   c. Now extend the looping algorithm to path finding through a Benes network. Estimate the time complexity for the algorithm for the Benes network.

3.11 References

The seminal work on multi-stage networks was done by Clos, who invented strict-sense non-blocking multi-stage networks. The unpublished work by Slepian (and a later published proof by Duguid) on non-blocking networks led to various optimal constructions of networks, such as the Benes network. The classical book by Benes, which collected a number of his important research results, stands as a treatise on the topology, algebra, and performance analysis of switching networks. Our treatment of strict-sense and rearrangeably nonblocking networks is a simplification of Paull's, which halved the number of circuit rearrangements in Slepian's result.

The next improvements on strict-sense non-blocking networks were made by Cantor, with a practical $4N(\log_2 N)^2$ complexity network, and by Bassalygo et. al., with an impractical but theoretically intriguing $CN \log_2 N$ complexity network. Our treatment of the Cantor network follows that of Masson et. al., which also gives an interesting tutorial of various kinds of interconnection networks.

The looping algorithm for the simultaneous setup of all connections was proposed by Opferman et. al. This algorithm can be parallelized using the method by Lev et. al.

Besides the design and control of interconnection networks, the operation and management of switching systems is a complicated subject beyond the scope of this book. A more complete picture of the design of commercial electronic switching systems (ESS) can be

3.11 References

found in two special issues of the Bell System Technical Journal, one by Spencer *et. al.* on the 4ESS, and another edited by Hayward on the 5ESS. A description of the ITT system 12 architecture can be found in the article by Bonami *et. al.*

1. L. A. Bassalygo and M. S. Pinsker, "On the complexity of optimal non-blocking switching networks without rearrangement," *Probl. Peredach. Inform.*, vol. 9, Jan. 1973.

2. V. E. Benes, *Mathematical Theory of Connecting Networks and Telephone Traffic*, New York: Academic, 1965.

3. R. Bonami, J. M. Cotton, and J. N. Denenberg: "ITT 1240 digital exchange: Architecture," *Electrical Communications*, vol 56, no 2/3, pp 126-134, 1981. Also the entire issue deals with the system 12 switch.

4. C. Clos, "A study of non-blocking switching networks," *Bell Syst. Tech. J.*, vol 32, pp. 406-424, March 1953

5. D. Cantor, "On non-blocking switching networks," *Networks*, vol. 1, Dec. 1971.

6. A. M. Duguid, "Structural properties of switching networks," Brown University, Progress Report BTL-7, 1959.

7. W. S. Hayward, editor, special issue on the 5ESS Switching System, *AT&T Technical Journal*, vol-64 no. 6, part 2, July-August 1985.

8. G. Lev, N. Pippenger, and L. G. Valiant, "A fast parallel algorithm for routing in permutation networks," IIEEE Trans. Computers, vol-30, no. 2, Feb. 1981.

9. G. M. Masson, G. G. Gingher, and S. Nakamura, "A sampler of circuit switching networks," *IEEE Computer*, June 1979.

10. D. C. Opferman, N. T. Tsao-Wu, "On a class of rearrangeable switching networks, part I: control algorithm," *Bell Syst. Tech. J.*, vol. 50, no. 5, May 1971.

11. M. C. Paull, "Reswitching of connection networks," *Bell Syst. Tech. J.*, vol 41, pp. 833-855, 1962.

12. D. Slepian, "Two theorems on a particular crossbar switching network," unpublished manuscript, 1952.

13. A. E. Spencer *et. al.*, Special issue on the 4ESS switch, *Bell Syst. Tech. J.* vol-61, no. 4, Sept. 1977.

CHAPTER 4

Multi-Point and Generalized Circuit Switching

- *4.1 Generalized Circuit Switching*
- *4.2 Combinatorial Bounds on Crosspoint Complexity*
- *4.3 Two Stage Factorization of Compact Superconcentrators*
- *4.4 Superconcentrators and Distribution Networks - A Dual*
- *4.5 Construction of Copy Networks*
- *4.6 Construction of Multi-Point Networks*
- *4.7 Alternative Multi-Point Networks*

4.1 Generalized Circuit Switching

In the previous chapter, we were concerned with making point-to-point connections. Formally, let I be a set of M inputs, and O be a set of N outputs. A point-to-point connection pattern is defined as a set of pairs $C = \{ (i,o) \}$ for which the $i \in I$ are distinct, and $o \in O$ are also distinct. In Mathematics, C is also known as a one-to-one mapping (figure 1).

In this chapter, we shall generalize the set of permissible connection patterns. The first generalization removes the constraint that the i of the pairs in C are distinct, while the o remain distinct. In other words, C is a one-to-many (figure 1) mapping for which we may connect an input to more than one output. We call this scenario multi-point switching or multi-casting.

Most generally, we may also consider the case for which any pair (i,o) is allowable. Of course, the pairs in C remain distinct.

These connection patterns can be conveniently expressed by graphs as shown in figure 1. We have the set of nodes I on one side and O on the other, with edges (i,o) linking the two sides. The prime objective of this chapter is the design and control of multi-stage interconnection networks for realizing multi-point connection patterns.

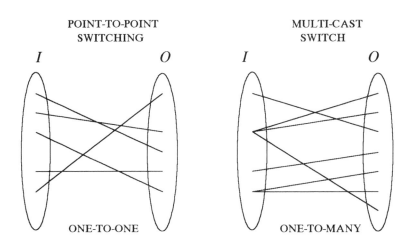

Figure 1. A Graph Representation of Connection Patterns

The multi-point multistage interconnection networks described later in the chapter make use of some output unspecific interconnection networks. An output unspecific network connects an input to some outputs, without regard to which specific outputs. We shall describe three such output unspecific connection functions: the concentrator, the superconcentrator, and the copy network as shown in figure 2. These networks are worth studying besides our concern for multi-point switching because they perform useful network functions as their names suggest. Let us now proceed to define these networks.

4.1 Generalized Circuit Switching

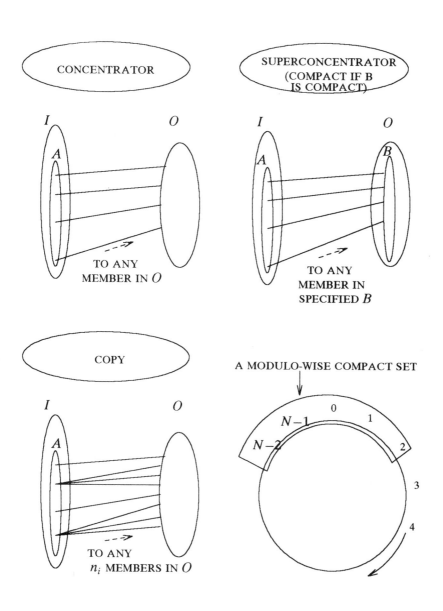

Figure 2. Connection Functions with Unspecified Output

For concentrators, a connection pattern C is defined by a subset A of I as shown in figure 2. Each member of A is to be connected to a distinct but unspecified $o \in O$.

For superconcentrators, we impose a slightly stronger condition on the set of connected outputs. A connection pattern C is defined by (A,B), A being a subset of I for which each member can be connected to a distinct but unspecified member in a subset B of O. We assume that the A and B specified are of the same size. Within the class of superconcentrators, we define a compact superconcentrator as one for which B is compact. A subset B of $\{0,1,2,\cdots,N-1\}$ is compact if the members of B can be arranged as a consecutive sequence modulo N. An example of a compact set is shown in figure 2.

For copy networks, a connection pattern C is defined by the set of pairs $\{\ (i,n_i)\ \}$ for which the input i can be connected to n_i unspecified outputs $o \in O$.

In order to establish the connection pattern for any one of the above functions, we need a graph G between the input links I and output links O. This graph, defined in general in the previous chapter, has nodes which are crosspoint switches. (An m-input n-output switch node has mn crosspoints.) By setting the crosspoints of the switch nodes, a connection is realized if there is a path from an input to an output.

Later in this chapter, we shall show that concentrators can be used to construct copy networks. In turn, a copy network (which makes output unspecific copies) cascaded with a point-to-point network (which makes output specific connections) constitutes a multi-point network. Before we focus on the construction of concentrators, the next section examines the crosspoint complexity (the crosspoint count) of these output specific and unspecific interconnection networks. In the process, we shall show that output specific networks must satisfy a complexity lower bound of $O(N\log N)$ and output unspecific networks must satisfy a complexity lower bound of $O(N)$. Throughout this book, we say a function $g(N)$ has a complexity $O(f(N))$ if $g(N)<\kappa f(N)$ for all $N>N_0$, for some constant κ and N_0.

4.2 Combinatorial Bounds on the Crosspoint Complexity

Let $\varsigma(G)$ be the logarithm (base 2) of the number of distinct and legitimate C realized by G. (Note that ς depends not only on G, but how we define the legitimate C for functions such as copying or multi-casting.) Effectively, ς measures the combinatorial power of a graph. However, ς may bear no direct relationship to the control complexity of finding a switch setting for realizing a connection pattern.

Let us denote an upper bound on $\varsigma(G)$ by $\overline{\varsigma}(G)$. An obvious upper bound is given by the total number, say R, of crosspoints in G. Each crosspoint can be set in two ways (connected or not connected), generating possibly two distinct C. There are at most 2^R crosspoint settings, hence we must have $\varsigma \leq R$.

Better upper bounds can be obtained through two observations. First, it is possible that some crosspoint settings may not produce a legitimate C for some functions. For example, we may have two crosspoints connected to the same output for point-to-point connections. Second, we may have 2 distinct crosspoint settings determining the same C. Better upper bounds can be obtained by eliminating these illegitimate or doubly counted connection patterns. However, such improvements often do not come easily.

For each connection function which defines the set of legitimate C, we may compute the logarithm of the total number of distinct C, denoted by $\underline{\varsigma}$. We say that a graph G is rearrangeably non-blocking if all such C can be realized by G. Obviously, we must have $\underline{\varsigma} \leq \varsigma(G)$ for rearrangeably non-blocking G.

In the remainder of this section, we shall compute $\underline{\varsigma}$ for various connection functions as a lower bound to the number of crosspoints for a rearrangeably non-blocking G.

1. Point-to-point connections

Let us assume that there are N elements in both I and O, and that each element in I is connected to a distinct element in O; such a connection pattern is termed maximal. Obviously, a network which can realize all maximal connection patterns can also realize less than maximal patterns.

Hence the number of C we want to realize is equal to $N!$. Using Sterling's approximation, we have

$$N! \approx \sqrt{2\pi} N^{N+\frac{1}{2}} e^{-N} = \sqrt{2\pi} \exp_2(N\log_2 N - N\log_2 e + \tfrac{1}{2}\log_2 N)$$

Therefore,

$$\zeta = \log_2 N! \approx N\log_2 N - 1.44N + \tfrac{1}{2}\log_2 N = O(N\log N)$$

Suppose we use 2×2 switching nodes which have two states, namely, the cross state and the pass state as shown in figure 3. We must have at least $N \log_2 N$ such nodes to realize the $N!$ possible maximal connection patterns. Hence a point-to-point interconnection network must have a node count of $O(N \log N)$.

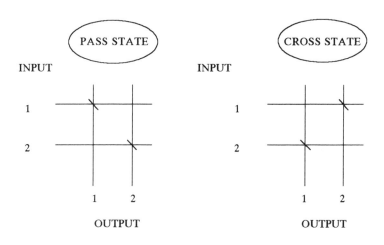

Figure 3. A 2 by 2 switching node with 2 states

Using the Benes network defined in the previous chapter, we have $2\log_2 N - 1$ stages each of $N/2$ switches of size 2×2. Hence for the Benes network, we have a total of $N\log_2 N - N/2$ binary switches. This number is very close to the lower bound ζ.

2. Multi-point switching

Let us assume that each output is connected to an input. In other words, each element in O may choose any one of the N inputs.

4.2 Combinatorial Bounds on the Crosspoint Complexity

Therefore, we have a total of $N^N = \exp_2(N\log_2 N)$ connection patterns C. Consequently, $\underline{\varsigma} = N\log_2 N$. Notice that this exponent is larger than the exponent of point-to-point switching by approximately $1.44N$.

In designing a multi-point network, this lower bound suggests that we should look for a network which has a switch node count roughly equal to $N\log N$, which is roughly N more than the node count for the Benes network. Unfortunately, no such network has been found to date. In this chapter, we shall describe a network which achieves multi-casting with twice the complexity of a Benes network.

Both point-to-point and multi-point interconnection networks have an $O(N\log N)$ complexity lower bound. In the following discussion, we shall show that the output unspecific interconnection networks, such as various types of concentrators and copy networks, has $O(N)$ (or linear) complexity.

3. Concentrators

For a concentrator with M inputs and N ($<M$) outputs, we defined a connection pattern C to be a set of any N of the M inputs. There are $\binom{M}{N}$ such sets. Hence using Sterling's approximation:

$$\underline{\varsigma} = \log_2 \frac{M!}{N!(M-N)!}$$

$$\approx \log_2 \sqrt{\frac{M}{2\pi N(M-N)}} \frac{M^M}{N^N(M-N)^{M-N}}$$

After some algebraic manipulation and ignoring lower order terms (namely the square roots), this can be expressed in terms of the binary entropy function H and the concentration ratio $c = N/M$:

$$\underline{\varsigma} \approx M\,H(c) \quad ; \quad H(c) \stackrel{\Delta}{=} -c\log_2 c - (1-c)\log_2(1-c)$$

For given c, we have $\underline{\varsigma} = O(M)$. This complexity lower bound is smaller by a factor of $\log M$ than that for point-to-point or multi-point networks.

Concentrators with linear complexity (namely that the crosspoint count is a constant times M) can be shown to exist, but the associated proportionality constant is prohibitively large compared with other practical constructions. Consequently, linear concentrators are more complex except for astronomical M and N. Not much is known concerning the control algorithm for these linear concentrators. We shall focus instead on constructions with $O(M\log M)$ complexity later in this chapter.

Aside from theoretical interest, there is very little practical incentive for constructing linear concentrators. The size (measured by M and N) of a concentrator is usually not too large in practice, in comparison to the size of a switch. Furthermore, if no rearrangement of connections is allowed (in other words, the concentrator is strict-sense non-blocking), it can be shown that such a concentrator must have an extra logarithmic factor in complexity.

We give here an intuitive argument explaining why strict-sense non-blocking concentrators are as complex as point-to-point non-blocking switching networks. On one hand, we can use a strict-sense non-blocking switching network as a strict-sense non-blocking concentrator. On the other hand, there are sequences of establishing and removing connections for a strict-sense non-blocking concentrator which force the concentrator to realize all possible point-to-point connection patterns as shown in the following argument. Suppose N out of M inputs are connected to all N outputs. We now remove any one of the connections. The next connection to be established, without choice since a rearrangement is not allowed, must use the recently vacated output. Continuing this alternating sequence of removing and establishing connections, we can obtain any point-to-point connection pattern. Hence a strict-sense non-blocking concentrator must have at least $O(N\log N)$ complexity.

4. Superconcentrators

It is obvious that superconcentrators and copy networks must have complexities at least as large as concentrators, since a subset of their connection patterns realizes the function of concentration.

Let us define the legitimate set of $C=(A,B)$ by all A and B with K elements, $K \leq M, N$, the number of inputs and outputs. Hence there are a total of $\binom{M}{K}\binom{N}{K}$ legitimate C. Hence

$$\underline{c} \approx M\, H(\frac{K}{M}) + N\, H(\frac{K}{N})$$

using the same approximation formula for combinatorial terms derived

4.2 Combinatorial Bounds on the Crosspoint Complexity

for concentrators. We notice readily that superconcentrators are more complex than concentrators by an extra linear (in N) term, provided that the concentration ratio K/M is kept constant.

For compact superconcentrators, the output set B is defined once the starting position of the compact sequence is specified. Since there are N possible starting positions, we have

$$\varsigma = M\, H(\frac{K}{M}) + \log_2 N$$

for compact superconcentrators is larger than that for concentrators by an additional term $\log_2 N$. Hence, a compact superconcentrator has a complexity lower bound close to that of a concentrator.

As we have shown before, a concentrator has a complexity lower bound roughly half that of a superconcentrator (assuming $M \approx N$). In fact, we may construct a superconcentrator from two compact superconcentrators in the following manner. We use two compact superconcentrators positioned back to back as shown in the figure 4. The second compact superconcentrator is used in a reversed fashion, namely that the inputs are used as outputs and vice versa. The set C in the middle is compact, while the set A and B are arbitrary. Hence this cascade is a superconcentrator. Given this method of constructing superconcentrators, we shall focus on how to construct compact superconcentrators in section 4.3.

5. Copy Networks

Assume that the sum of the number of copy requests n_i over all inputs i is equal to N, the number of outputs. The number of distinct C is given by $\binom{M-1+N}{M-1}$, which is the number of ways of putting N objects in M bins. Consequently

$$\varsigma \approx (M-1+N)\, H(\frac{M-1}{M-1+N})$$

Again, the complexity lower bound is linear in M and N.

6. Many-to-Many Networks

Consider the most general case of allowing any (i,o) in the connection pattern C. There are N^2 distinct (i,o). Accounting for the inclusion or exclusion of each distinct (i,o) in C, we have a total

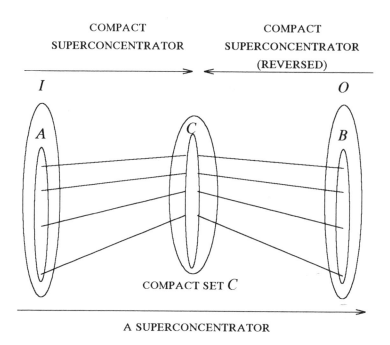

Figure 4. Constructing Superconcentrators Using Compact Superconcentrators

of $\exp_2(N^2)$ distinct C. Consequently, $\varsigma = N^2$. This complexity lower bound is a significant jump from the $O(N\log N)$ complexity for point-to-point or multi-point connection patterns.

4.3 Two Stage Factorization of Compact Superconcentrators

After examining the combinatorial lower bound for the complexity of these generalized switching networks, we now study specific constructions for these networks. In this section, we construct $M \times N$ compact superconcentrators. An obvious construction for superconcentrators is an $M \times N$ crossbar switch. In many applications for which M is not very large, such a single stage construction is often

4.3 Two Stage Factorization of Compact Superconcentrators

employed. For larger M and N, we use a two stage recursive factorization process for constructing compact superconcentrators with $O(M\log M)$ crosspoint complexity. This factorization process is also crucial to the discussion of multi-point and copy networks in later sections.

Let us assume that p divides M and q divides N. Let us construct a 3 dimensional structure with 2 stages as shown in figure 5. The first stage has p horizontal planes each of which is an M/p-input q-output compact superconcentrator. The second stage has q vertical planes each of which is a p-input N/q-output compact superconcentrator.

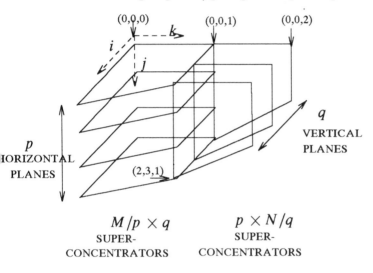

Figure 5. Two Stage Factorization for Compact Superconcentrators

In figure 5, a link is represented by the intersecting point between two perpendicular planes in successive stages. Let us represent the output i of plane j at stage k by its Cartesian coordinates (i,j,k). The inputs to the network have coordinates $(i,j,0)$ and the outputs have coordinates $(i,j,2)$. For a given k, the coordinates (i,j,k) are ordered in the row major sense, namely $(i,j,k)>(i'j',k)$ if $j>j'$. If $j=j'$, then $(i,j,k)>(i',j',k)$ if $i>i'$.

The overall concentration ratio for this two stage structure is given by $c=N/M$. By this two stage factorization, we factor c into a concentration ratio of $c_1 = pq/M$ for the first stage, and a concentration ratio of $c_2 = N/pq$ for the second stage.

For a given c, how should the concentration ratios c_1 and c_2 be assigned for the two stages? This depends on how much blocking (the event that an active input cannot be connected to an output) should occur in each stage. Let us assume that at most N of the M inputs are active (otherwise, any $M \times N$ concentrator would become blocking). In order that the first stage concentrators are not blocking, we must either have q (the number of outputs per first stage concentrator) be equal to N, or have $c_1 = 1$, in which case $q = M/p$. In the case of $q = N$, a second stage concentrator has only 1 output since $N/q = 1$. In the case of $q = M/p$, the first stage concentrator cannot be blocking since it does not really concentrate. However, it performs the function of compact superconcentration, which is necessary as we shall show later.

We now show that the two stage structure is indeed a $M \times N$ compact superconcentrator. Consider first the special case of $q = N$. Let $M = pN$ where p is an integer greater than 1. The resulting compact superconcentrator is shown in figure 6. The concentration ratio c is given by $1/p$, which is achieved in the second stage.

This two stage structure is a compact superconcentrator by the following output assignment algorithm. The first horizontal plane of the first stage uses up the smaller outputs in a compact manner, then the second horizontal plane is allowed to use the next smaller outputs, etc. The resulting output assignment for all horizontal planes is also compact.

We now consider the case when the second stage compact superconcentrators each has more than one output. (For example, when $q = N/2$. The first stage compact superconcentrators must have $c_1 = 1$ as we argued before). In other words, there is more than one row at the output of the second stage. We shall show that the two stage structure compact superconcentrates if each plane in the two stages compact superconcentrates. The proof consists of stating a specific manner of output assignment. We use the output rows in a row major fashion for the first horizontal plane, the second horizontal plane, etc.

Row Major Assignment Algorithm for Compact Superconcentrators:

For the two stage factorization shown in figure 5, an active input $(i, j, 0)$ is connected to an output of the compact superconcentrator

4.3 Two Stage Factorization of Compact Superconcentrators

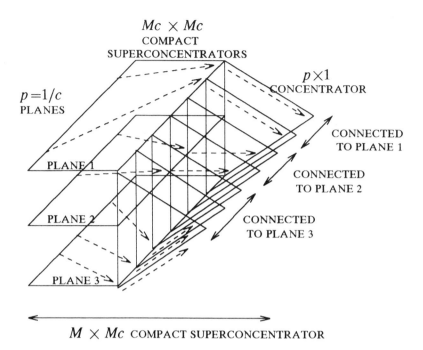

Figure 6. Connection Assignment for Multi-Stage Compact Superconcentrators

by the following algorithm:

1. Initialize the links being considered:
   The input $s_0=(i,j,0)=(0,0,0)$
   The first stage output $s_1=(i',j,1)=(0,0,1)$
   The second stage output $s_2=(i',j',2)=(0,0,2)$.

2. If s_0 is active, then:
   - 2.1 Connect s_0 to s_1, and connect s_1 to s_2.
   - 2.2 $i' \leftarrow i'+1$. If $i'=q$, then $i' \leftarrow 0$ and $j' \leftarrow j'+1$.

3. $i \leftarrow i+1$. If $i=M/p$, then $i \leftarrow 0$ and $j \leftarrow j+1$.

4. Go to 2 unless we come to the last input s_0.

Notice that i' is updated only for active inputs, hence i' increases more slowly than i. Therefore, there is always an output s_1 of the first stage concentrator available for each active input. Furthermore, any initial output other than $s_2=(0,0,2)$ may be used provided the increments are made in a row major modulo manner.

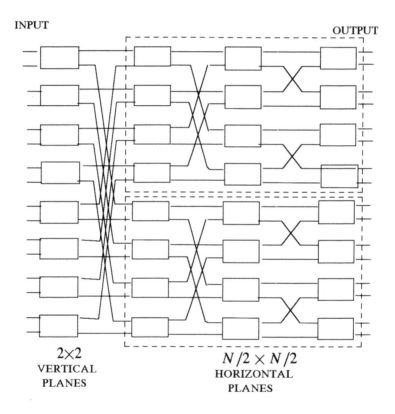

Figure 7. An Inverse Banyan Network Formed by Recursive 2-Stage Factoring

We have shown that the two stage factorization indeed produces a compact superconcentrator using the row major assignment algorithm. This two stage factorization procedure can be applied recursively to the compact superconcentrators in either stage. Consider $M=N=2^n$. Repeated factorization (with $p=N/2$ and $q=2$) generates a network of 2×2 compact superconcentrators, which can be realized by 2×2 switch nodes. The resulting network shown in figure 7 is called an inverse banyan network. Figure 7, which is drawn in two dimensions, is a flattened version of the three dimensional representation of figure 5, with the order of the switches in each stage preserved. (The first stage 2×2 switches of figure 7 are the horizontal first stage switches in figure 5. The dotted squares in figure 7 are the vertical second stage $\frac{N}{2} \times \frac{N}{2}$ concentrators in figure 5.) We leave this mental exercise to the readers. A banyan network is the mirror image of the inverse banyan network by reversing the front and back ends of the network.

4.4 Superconcentrators and Distribution Networks - A Dual

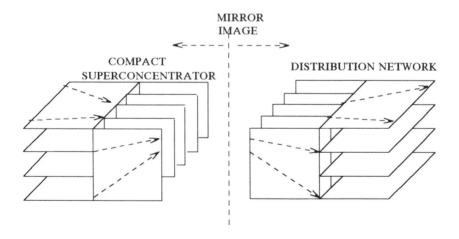

Figure 8. Construction of a Distribution Network

After showing that the two stage factorization can be used for constructing compact superconcentrators, we now show another use for this two stage construction. Under some special conditions on a connection pattern, the compact superconcentrator generated by the two stage factorization process can be used for point-to-point switching, as stated in the following theorem.

Theorem:

Consider the 2 stage network, called a distribution network, given by the mirror image of the compact superconcentrator as shown in figure 8.

Suppose the input-output connection pattern $C=\{(i,o_i)\}$ satisfies the following conditions:

1. (Compactness Condition)
   The active inputs i for the pairs in C are compact in a modulo fashion.

2. (Monotone Condition)
   The output o_i to be connected to each active input are strictly increasing in i in a modulo fashion as shown in figure 9.

All point-to-point connection patterns satisfying these two conditions can be connected using the distribution network.

Proof:

Intuitively, the distribution network with input-output connection patterns satisfying these two conditions performs the inverse function of a compact superconcentrator, namely that it distributes instead of concentrates as shown in figure 8. Let us treat the outputs of the superconcentrator as inputs to the distribution network. Likewise, we treat the inputs of the superconcentrator as outputs of the distribution network. The compactness of the inputs to a distribution network is a dual aspect of the compacting function of the superconcentrator. The monotonicity of the outputs to a distribution network is a dual aspect of the row major assignment algorithm. □

This theorem states that we may make a two stage network non-blocking if the connection requests arrive in sorted order. One way to achieve this is to put a sorting network in front of the two stage network, which shall be treated in more detail in chapter 6.

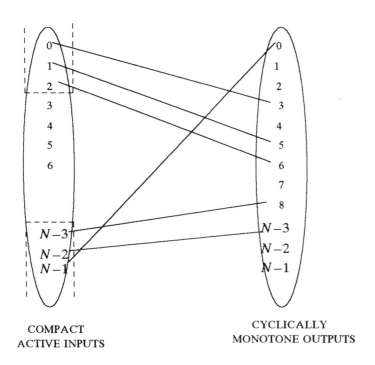

Figure 9. A Compact and Monotone Connection Pattern

4.5 Construction of Copy Networks

The distribution network of figure 8 can be modified to perform the function of copying. The inputs to the switch planes are now allowed to make multi-point connection to the outputs. The distribution network which allows multiple connections between an input and outputs is called a copy distribution network. As shown in figure 10, an input can make extra connections to outputs provided that the outputs connected remain monotonically increasing with respect to the inputs in a manner to be defined later.

Let us state this copy mechanism more formally. As defined before, a connection pattern for a copy network is given by $C = \{(i, n_i)\}$, for which n_i is the number of copies to be made. Suppose in a realization of the copy function, we have $C = \{(i, O_i)\}$ in which

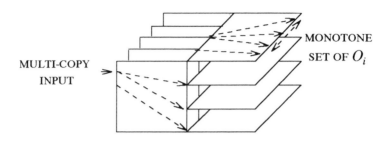

Figure 10. A Copy Distribution Network

O_i is the set of n_i outputs assigned to input i. We want to show that the distribution network also suffices for copying if C satisfies the following conditions.

1. (Compactness Condition)
   Same as before, the active inputs i are compact in a modulo fashion.

2. (Monotone Condition)
   Each element in O_i is greater than each element in $O_{i'}$ if $i > i'$, modulo-wise as shown on the right hand side of figure 11.

As shown in the previous section, there is a dual relationship between distribution and concentration. Likewise, there is a dual relationship between copy distribution and a new connection pattern called "many-to-one concentration" defined as follows. The connection patterns $C = \{(i,o)\}$ concentrates in a many-to-one manner when consecutive active inputs could be connected to the same output. A many-to-one concentration pattern is shown on the left hand side of figure 11.

Many-to-one compact superconcentration can also be realized via the two stage factorization method described for ordinary compact superconcentration. The only difference is that two connections (i,o), (i',o) for the same output and with i, i' on the same horizontal plane in the first stage may use the same link between the

4.5 Construction of Copy Networks

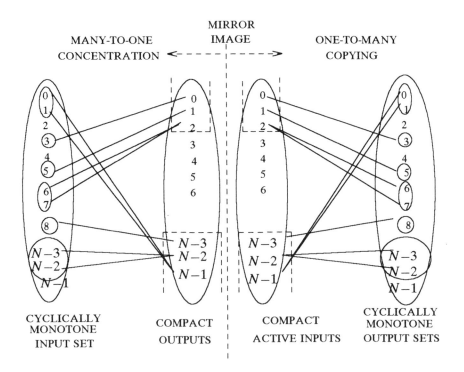

Figure 11. Duals: Many-to-One Concentration and Copying

first and second stage. Thus the row major assignment algorithm described is modified accordingly.

The inverse of a many-to-one concentrator performs the copy function, with an input connected to possibly more than one output. The copy distribution network of figure 10 is not yet a copy network for two reasons. First, the active inputs to a copy network are not necessarily compact as required for a copy distribution network. Hence we have to concentrate the active inputs so that they become compact. This is easily achieved by using a compact superconcentrator first, as shown in figure 12. Second, we have to assign O_i for each active input i in order to satisfy the monotone condition. There is considerable flexibility for this assignment since it is quite immaterial where the copies should appear at the output of

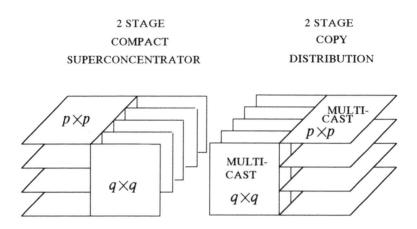

Figure 12. Copy=Superconcentrator+Inverse-Superconcentrator

the copy network. One convenient way of mapping the inputs i to sets of monotone O_i is by specifying an interval of outputs of size n_i (the number of copies to be made) with the two boundaries of the interval given by the total number of copies requested by all $i'<i$ and $i'\leq i$ respectively. In other words, $O_i = [\sum_{i'<i} n_{i'}, \sum_{i'\leq i} n_{i'}-1]$. For $n_i=0$, the interval is an empty set.

4.6 Construction of Multi-Point Networks

One method of constructing a multi-point network is by concatenating a copy network with a point-to-point network. As we have shown in the previous section, a copy network comprises a superconcentrator and a copy distribution network, which is the dual of the superconcentrator. We can use the three stage factorization process of the previous chapter for point-to-point switching.

By recursive factorization, the preceding cascade of networks consists then of the following sequence of networks: concentration, copy distribution, and the Benes network. The total number of 2×2 switching nodes in these networks is roughly $2N\log N$, which is twice as complex as a Benes network.

4.6 Construction of Multi-Point Networks

Suppose we perform a two stage factorization only once for the above functions of concentration and copy distribution, and a three stage factorization for the point-to-point switch. As shown in figure 13, we obtain a cascade of a 2 stage superconcentrator, a 2 stage copy distribution network, and a 3 stage Clos network. There are altogether 7 stages, with the third and fourth stages requiring multi-point switches.

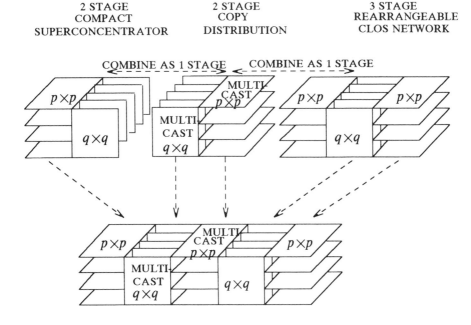

Figure 13. A 5 Stage Multi-Point Network

Examining these 7 stages more closely, we observe that only five stages are actually needed. Since the superconcentrator is the inverse of the copy distribution network, the second and third stage switch planes are parallel. Therefore, we can combine the switch planes of these two stages with a single stage of vertical switch planes each with a multi-point switch. Likewise, the second stage of the copy distribution network and the first stage of the Clos network both have horizontal switch planes, and therefore can be combined into a single stage of horizontal switch planes each with a multi-point switch.

So far, we have not discussed about the control complexity of the above networks. It is obvious that these networks are not strict-sense non-blocking. However, there are efficient parallel algorithms for path setup for the superconcentrator and the copy distribution network. Hence this multi-point network can be adapted for packet switching. We defer the discussion to Appendix 6A which describes multi-point packet switching.

4.7 Alternative Multi-Point Networks

There are multi-point connection networks which do not factor the multi-point connection function into the cascaded functions of copying and point-to-point switching. However, their complexity growth rate is more than $O(N\log N)$.

One such construction of an $N \times N$ multi-point connection network was proposed by Pippenger as shown in figure 14. A 1×2 demultiplexer connects each input to two concentrators, each of size $N \times \frac{1}{2}N$. The outputs of each concentrator are then connected to an $\frac{1}{2}N \times \frac{1}{2}N$ multi-point connection network. The two half-size multi-point connection networks can be recursively constructed by the same procedure.

This construction indeed realizes all multi-point connection patterns. If an input is to be connected to outputs of both the upper half and the lower half of the N outputs, the 1×2 demultiplexer would connect the input to both halves of the network. There can be at most $\frac{1}{2}N$ active inputs to each concentrator, which concentrates these inputs to the proceeding $\frac{1}{2}N \times \frac{1}{2}N$ multi-point network.

If the concentrators used are strict-sense non-blocking, then the resulting multi-point network is also strict-sense non-blocking. Unfortunately, little is known about practical and efficient strict-sense non-blocking concentrators. Let us employ instead the rearrangeable concentrators constructed by two stage factorization. The $N \times \frac{1}{2}N$ concentrator can be constructed in the manner shown in figure 6, namely with $p=2$ horizontal planes of $\frac{1}{2}N \times \frac{1}{2}N$ concentrators, and the concentration factor $c=\frac{1}{2}$ is achieved by the 2×1 vertical concentrators. We may use an inverse banyan network as an $\frac{1}{2}N \times \frac{1}{2}N$ concentrator. Let us now estimate the complexity of the

4.7 Alternative Multi-Point Networks

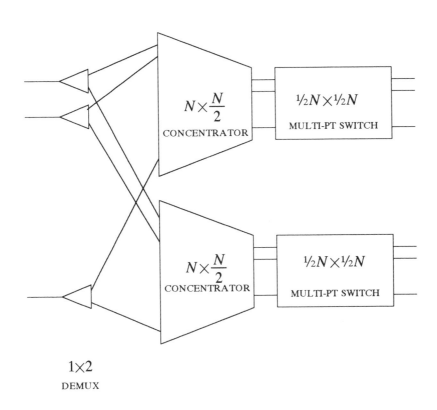

Figure 14. The $N \times N$ Multi-Point Pippenger Network

resulting multi-point connection network, focusing on the complexity of the constituent banyan networks and ignoring the complexity of the 1×2 demultiplexers as well as the 2×1 concentrators. There are 4 banyan networks each with $(N/4) \log_2(N/2)$ 2×2 nodes, 8 banyan networks each with $(N/8) \log_2(N/4)$ nodes, etc.; resulting in a total complexity of roughly $\frac{1}{2}N(\log_2 N)^2$.

Besides the Pippenger network, let us consider briefly how we may use the 3 stage Clos network of the previous chapter for multi-point switching. If we use Paull's matrix for representing paths for multi-point connections (see section 3.4), we have the following conditions for a legitimate multi-point connection matrix:

- Each column can have at most $\min(r_2, n_3)$ distinct symbols representing the middle stage switch being connected to a third stage switch.
- There is no restriction on how many symbols or whether they are distinct for entries in a row. Repeated symbols in the same row represent a multi-point branching of an input in the middle stage switch represented by the symbol.

It is known that a large number of middle stage switches are needed for making the Clos network both multi-point and non-blocking. Nevertheless, the Clos network with a moderate number of middle stage switches may be used for multi-point switching, if we tolerate a certain probability of blocking.

The Clos network has been modified by various means for non-blocking and multi-point switching. A description of these networks

4.8 Exercises

1. Consider the Bell System number 5 crossbar network with a network structure shown in figure 15. It is constructed using $n \times n$ switches. A switch frame (in dotted boxes) is an $n^2 \times n^2$ network with 2 stages with n switches in each stage. The entire network has 2 stages of n frames, with n^3 inputs and n^3 outputs.

    a. Compute the number of crosspoints for the number 5 crossbar. For $n=10$, compare that with the number of crosspoints needed by a 3 stage rearrangeably non-blocking network and the Benes network. (Use the approximation $2^{10} \approx 1000$. For the 3 stage network, assume there are 32 32×32 switches in each of the 3 stages.)

    b. How many paths are there between an input and an output?

    c. Show that the number of permutations realized by the network is upper bounded by $(n!)^{4n^2}$. Why is this only an upper bound?

    d. Upper bound the fraction of permutations which can be realized by the network. Get a rough estimate for $n=10$ by using Sterling's approximation

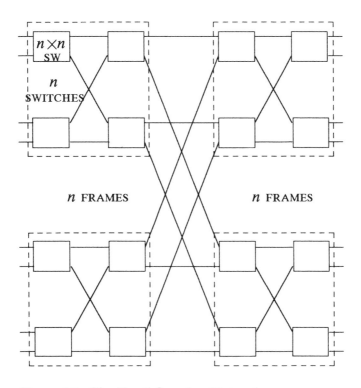

Figure 15. The No. 5 Crossbar Network

$$\log_{10} x! \approx (x+\frac{1}{2})\log_{10} x + x\log_{10} e$$

$$= (x+0.5)\log_{10} x + 0.4343x$$

e. Does the fraction surprise you? In order to reduce blocking, let us reduce the traffic to the network. Instead of a full permutation, let us assume only a fraction f of the inputs request a connection to distinct outputs. Get a rough estimate of f in order that the reduced number of permutation requests is comparable to the upper bound given in c. Explain why the number 5 crossbar is not bad at all in practice, using the statistics given in the exercise in chapter 1.

2. A 2×2 crossbar has 4 crosspoints, giving a total of $2^4=16$ crosspoint settings. Which of these constitute legitimate point-to-point or multi-cast connection patterns?

3. Why isn't a two stage structure of superconcentrators a superconcentrator, unlike the case for compact superconcentrators? (Hint: Construct a counter example.)

4. Construct a 128×8 concentrator using 8×8 crossbars and 4×1 multiplexers. The concentrator should be non-blocking in the sense that any 8 out of the 128 inputs can be connected to the 8 outputs.

5. Consider an extension of the Pippenger multi-point network.

   a. Figure 6 shows a way to construct an $N \times \frac{N}{p}$ concentrator with $p>1$ being an integer. Show that the crosspoint count of this construction is $N+pC(N/p)$, where $C(N)$ is the complexity of the $N \times N$ compact superconcentrator.

   b. Now replace the 1×2 demultiplexers in figure 14 by 1×p demultiplexers, the $N \times \frac{N}{2}$ concentrators by $N \times \frac{N}{p}$ concentrators, and the 2 $\frac{N}{2} \times \frac{N}{2}$ multi-point switches by p $\frac{N}{p} \times \frac{N}{p}$ multi-point switches. For linear complexity concentrators, we have $C(N) = \kappa N$. Using recursion and assuming that $N=p^n$ is large, show that the crosspoint complexity of the multi-point network is

   $$F(N) \approx (2+\kappa)pN\log_p N$$

   c. Alternatively, let us build concentrators from a recursive two stage factoring using $p \times p$ crossbars. Consequently, $C(N) = pN\log_p N$. We assume $N=p^n$, hence $C(p^n) = p^{n+1}n$. Show that for large N, the crosspoint complexity of the Pippenger multi-cast network is

   $$F(N) \approx p^2 N \frac{(\log_p N)^2}{2}$$

d. For the above expression, show that $p=3$ minimizes $F(N)$.

4.9 References

Generalized interconnection networks are not as well understood as point-to-point interconnection networks. Unlike point-to-point networks, practical constructions of copy networks and concentration networks exceed the complexity order of the theoretical lower bound. More results are expected in the future. However, the difficulty could also be intrinsic, as suggested by Pippenger on the complexity of strictly nonblocking concentration network. Our proof is a simpler version of Pippenger's.

The $O(N \log N)$ concentration, distribution and copy networks as well as their use for constructing generalized interconnection networks were first proposed by Ofman, and described further in an article by Thompson. The bulk of this chapter expands the treatment for this approach.

The existence of linear concentration and distribution networks were shown by Pinsker using a counting argument. An explicit but impractical construction was given by Margulis.

Pippenger's doctoral dissertation is the first comprehensive treatment on generalized interconnection network. The $O(N(\log N)^2)$ complexity multi-point connection network of section 4.7 is described in his dissertation. The excessive number of middle stage switches for making the three stage Clos network multi-point and rearrangeable was shown by Hwang. There are other practical generalized interconnection networks with a complexity growth higher than that proposed by Ofman. The treatment of these networks proposed by Masson *et. al.* and Richards *et. al.* is beyond the scope of this book.

1. F. K. Hwang, "Rearrangeability of multiconnection three-stage networks," *Networks*, vol. 2, pp. 301-306, 1972.

2. G. A. Margulis, "Explicit constructions of concentrations," *Problemy Peredachi Informatsii* vol. 9 no. 4, 1973. (English translation in: *Problems of Information Translation*, Plenum. New York, 1975)

3. G. M. Masson and B. W. Jordan, Jr., "Generalized multistage connection networks," *Networks*, vol. 2, pp. 191-209, 1972.

4. J. P. Ofman, "A universal automaton," *Trans. Moscow Math. Soc.*, vol. 14, 1975 (translation published by Amer. Math. Soc., Providence, RI, 1967, pp. 200-215.)

5. M. S. Pinsker, "On the complexity of a concentrator," *Seventh International Teletraffic Congress*, Stockholm, 1973.

6. N. Pippenger, "On the complexity of strictly nonblocking concentration networks," *IEEE Trans. on Comm.* , Nov. 1974.

7. N. Pippenger, "The complexity theory of switching networks," Ph. D. thesis, also MIT Res. Lab. of Electronics, Rep. TR-487, 1973.

8. G. W. Richards and F. K. Hwang, "A two-stage rearrangeable broadcast switching network," *IEEE Trans. Commun.*, vol-33, pp. 1025-1035, 1985.

9. C. D. Thompson, "Generalized connection networks for parallel processor interconnection," *IEEE Trans. on Computers*, vol-27, no 12, Dec. 1978.

CHAPTER 5

From Multi-Rate Circuit Switching to Fast Packet Switching

- 5.1 Depth-First-Search Circuit Hunting
- 5.2 Point-to-Point Interconnection of Multi-Slot TDM
- 5.3 Fast Packet Switching
- 5.4 Self-Routing Banyan Networks
- 5.5 Combinatorial Limitations of Banyan Networks
- 5.6 Appendix-Non-Blocking Conditions for Banyan Networks

In practice, many large capacity switching systems have switching networks with more than three stages. The complexity of the control algorithm is a major bottleneck on the rate at which circuits can be established. The circuit rearrangement algorithms studied in chapter 3 require excessive computation. Circuit rearrangement algorithms are seldom used in practice. Instead a depth-first-search circuit hunting algorithm is usually employed. This is described in section 5.1.

Two new developments in providing services increase the strain on the process of circuit hunting. First, the deployment of fast circuit switching (described in section 2.4) for bursty services may increase the rate at which circuit setups are requested. Second, the integration of multi-rate services in a switching network requires the control algorithm to be capable of hunting for multiple slots for a service. This problem of multi-slot connection setup is described in section 5.2.

To alleviate this control bottleneck, fast packet switching techniques are often used for multi-rate and bursty services. This is described in section 5.3.

In particular, the banyan class of interconnection networks is often used for fast packet switching. (Through this chapter, the name banyan network represents not only the banyan network itself, but all essentially equivalent networks, such as the baseline network defined in chapter 3. These networks have the advantage of self-routing a packet through the switching network. Some properties of these networks are described in sections 5.4 and 5.5.

5.1 Depth-First-Search Circuit Hunting

We want to hunt for a free path in a switching network G from an input to a desired output. Using depth first search, we search as far as possible down a free path which may lead to the desired output, and backtrack when the switching node visited has no free outgoing path to the output, or if the switching node has been visited before. Consider the multi-stage network shown in figure 1, for which the edges are denoted by the triplet (i,j,k) representing the outgoing edge i of the switch node j in stage k. (The indices i and j start from 0.) Suppose we want to connect input $s_t=(i_t,j_t,0)$ to the output $s_r=(i_r,j_r,K)$ in this network with K stages. We use the following algorithm for path setup.

Depth-First-Search Control Algorithm

1. (Initialization)
   $s=(i,j,k)\leftarrow(0,j_t,0)$.
2. (Termination)
   If $s=s_r$, stop.
3. (Depth Search)
   If s is free, and s may reach s_r through switch j' in the next

5.1 Depth-First-Search Circuit Hunting

stage, and j' has not been visited before, then $s \leftarrow (i=0, j', k+1)$ and go to step 2.

4. (Breadth Search)
   Otherwise $s \leftarrow (i+1, j, k)$ and go to step 2 provided that $i+1$ is not equal to the number of outgoing edges.

5. (Backtrack)
   If $i+1$ is equal to the number of outgoing edges, then backtrack to the switch node searched most recently in the previous stage. Assume that this backtrack is made along an edge represented by (i, j, k). Update $s \leftarrow (i+1, j, k)$. Go to step 2 if $i+1$ is not equal to the number of outgoing edges. Otherwise, backtrack further by repeating this step. If by backtracking we return to s_t, requested connection is blocked. Stop.

Let us also examine the search algorithms of some switching systems developed.

1. The search algorithm for the earliest automatic switching systems (called step-by-step switching) does not backtrack. It actually sets up the connection path as the search moves forward. The connection is blocked whenever the edge searched is not free.

2. In some switching systems with programmed control (say the 4ESS cited in the references of chapter 3), the connection state of the network is represented by a memory map in a central processor. The signal path is set up in the switching network after a free path is found in the memory map.

3. In systems with distributed path searching (say the ITT system 12 cited in the references of chapter 3), the switch nodes hold the local connection information. Path hunting is made from node to node, similar to step-by-step switching. However, backtracking is allowed. The path hunting is considered distributed because many connection requests can search for a free path simultaneously.

4. For the breadth search portion of the depth first search algorithm, the choice of the next edge for searching can be random among all outgoing edges instead of sequential (with the update of $i+1$ in the original algorithm). There is a tradeoff between random search versus sequential search. Using the same search order for every connection request, sequential search packs paths together by using heavily loaded switch nodes first before considering lightly loaded switch nodes. This packing

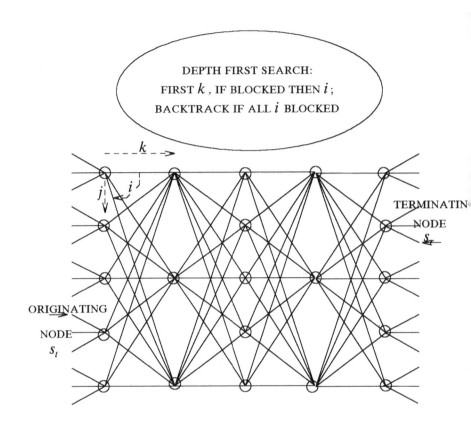

Figure 1. Depth-First-Search for Circuit Hunting

behavior reduces the probability of blocking connection requests. On the other hand, random search distributes the paths evenly in the entire network. The advantage of random search is that it often reduces the mean number of searches before a free path is discovered.

It should be emphasized that in practice, the search algorithm, more often than the network itself, causes blocking of connection requests. Exhaustive search of all possible paths is often impractical, hence most algorithms give up after a certain number of trials. The number of paths searched per connection request depends strongly on

5.1 Depth-First-Search Circuit Hunting

the load of the network. To reduce the number of search per connection request, the load is often reduced with respect to the theoretical maximum of allowing circuit rearrangement and unlimited search.

5.2 Point-to-Point Interconnection of Multi-Slot TDM

At the end of chapter 2, we showed that time slot interchanged TDM links interconnected by a TMS constitute TST multi-stage switching. We also showed that a TST switch is mathematically equivalent to SSS multi-stage switching. Let us generalize the problem to that of interconnecting multi-slot TDM links in a multi-rate service environment.

For multi-slot TDM, a terminal can be assigned more than one slot within the TDM frame. The information from the terminal is conveyed by these slots, which are time slot interchanged before being switched by the TMS.

It is apparent that we can use the same switching techniques in chapter 3 for establishing a multi-slot connection. Nothing is really changed except a connection occupying several slots could be viewed as several slot connections between the input and the output of the TMS. We also observe that the input-output connection pattern across the TMS can be different for each time slot as shown in figure 2. If the TMS is a multi-stage network, the paths through the TMS can be different for each slot. In particular, let the TMS be a 3 stage SSS network. The overall network becomes a 5 stage TSSST network.

Consequently, the time slots of a multi-slot connection use different paths in the TMS. Each path requires a setup, hence the complexity of setting up a multi-slot connection is proportional to the number of slots required. Since the number of slots in a frame can be quite large, the complexity of setting up a high bandwidth connection can also be very large.

To tackle this problem, we may restrict each slot in a multi-slot connection to use the same path in the TMS. This leads to the problem of slot mismatch as follows. Consider a network G for which the edges are time slotted transmission links. Consider a path P, which is a set of connected links, used for establishing a new connection. For each edge $l \epsilon P$, we have a frame occupancy vector which denotes slot occupancies in a frame. We may establish a

118   5. From Multi-Rate Circuit Switching to Fast Packet Switching

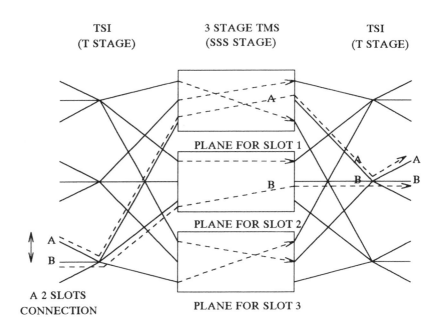

Figure 2. A Multi-Slot Connection

connection using the common vacant slots for the edges in the path. Most likely, few commonly vacant slots would be available for the new connection, because the vacant slots for different edges may be mismatched in their positions.

This mismatch problem can be avoided if a time slot interchanger (TSI) is used at the input or the output of each switch node. The non-blocking conditions are given by the theory developed in chapter 3. Thus each switch node, instead of a single S switch, becomes either a TS switch node or an ST switch node, depending on whether the TSI at the input or the output of a switch node.

Suppose we build a 3 stage switching network, for which each switching node in a stage is a TS switch node. For this 3 stage TS-TS-TS switching network, we can still use Paull's connection matrix for registering multi-rate connections. Instead of the condition that all symbols in a row (or column) are distinct, we may allow fractional

5.2 Point-to-Point Interconnection of Multi-Slot TDM

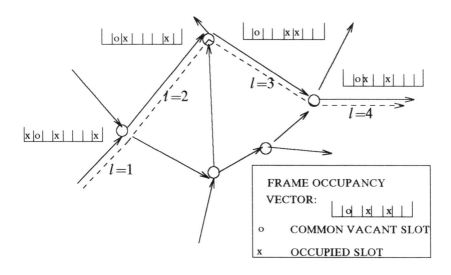

Figure 3. Time Slot Mismatch Between Edges in a Network

entries as long as the row (or column) sum is at most 1.

We can also use the depth-first algorithm for finding a path with sufficient capacity for a connection. Backtracking occurs when the edge considered does not have sufficient capacity. Obviously, a connection requiring higher bandwidth suffers a higher probability of backtracking. Consequently, one major disadvantage of multi-slot circuit switching is that it tends to penalize higher bandwidth connections with substantially higher probability of blocking.

5.3 Fast Packet Switching

In chapter 2, we arrived at the conclusion that packet multiplexing is more flexible than multi-slot TDM for multiplexing multi-rate connections. For switching multi-rate connections, we draw a similar conclusion favoring packet switching over multi-slot circuit switching for the following reasons:

1. By using a first come first serve queueing discipline, we avoid the need to keep track of slot assignments in a frame. Slot assignment across a switching network is computationally intensive, and more so when the connection has fluctuating bit rate.
2. We no longer need a large memory for the entire frame, but rather a smaller memory for queueing packets.

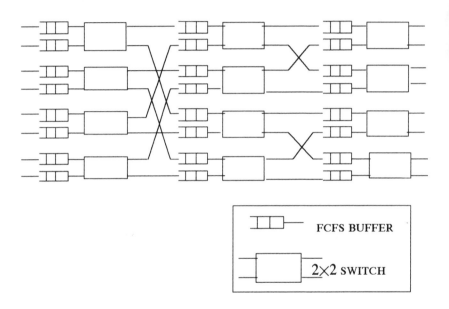

Figure 4. A Fast Packet Switch

Structurally, a multi-stage TS circuit switch described in the preceding section is similar to a fast packet switch: we simply convert the T switch from a TSI into a First Come First Served (FCFS) queue at each input as shown in figure 4. The packet switch uses input buffering at each switch node.

A fast packet switch may transport either fixed length packets or variable length packets. The switch node connects one of the input queues to an output for transmitting the packet at the head of that input queue. Upon completion of transmission of a packet, that

5.3 Fast Packet Switching

output is reconnected to any input waiting for service for that output. In this manner, control of the switch node depends only on local information, unlike multi-slot circuit switching which requires slot matching across the entire network.

The fast packet switch shown in figure 4 uses the banyan network introduced in the previous chapters. For this network, there is a unique path between each input and output of the network. We also call this packet switch a buffered banyan network.

Since service of a packet depends on the local contention for the output, there is a random delay before a packet is served at a switch node. Coupled with the random arrivals of packets, this random delay may cause buffer overflow occasionally. A packet arriving at a full input buffer is lost. A mechanism for avoiding packet loss is needed, as shown in figure 5. A full buffer informs the upstream switch nodes connected to the buffer, to prevent these nodes from transmitting to the full buffer. This feedback exerts a back pressure to the flow of packets in a network. Therefore, whether a packet at the head of an input queue is served depends on two factor: first, output contention among the conflicting inputs; and second, the availability of buffer space downstream.

There are many variations and improvements for this fast packet switch. We shall discuss a few of them.

1. Virtual cut through

Upon the arrival of a packet at a node, we may immediately transmit the packet on the desired outgoing link if the link is free, thereby avoiding buffering at the node. This cut through mechanism is illustrated in figure 5. A packet is buffered only if the outgoing link is not free or the buffer fed by the outgoing link is full. This cut through process reduces delay and improves throughput.

2. Output buffering

For input buffering, only one of the packets from different inputs contending for the same output is delivered. Alternatively, a switch node can be built such that all contending packets at different inputs are delivered and stored at the outputs. Output buffering thereby gives a better throughput than input buffering. However, output buffering requires each switch node to be capable of simultaneously connecting all its inputs to an output. Also the output buffer must be able to receive and manage many packets simultaneously. Figure 6 shows one

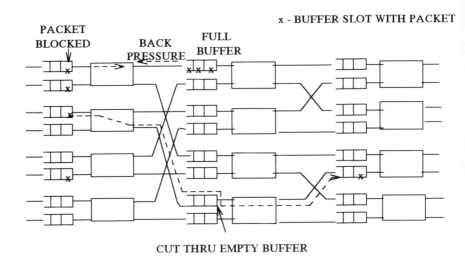

Figure 5. Virtual Cut Throughput of a Switch Node

such implementation using multiple input buses, each tapped by all outputs. The complexity of this scheme can be excessive if the switch node has many inputs.

Output buffering can also be effectively implemented by a very fast memory as shown in figure 7. Each switching node within the multistage switching network is implemented by a common memory that is capable of reading all packets from the inputs to the node in a slot time (assuming fixed length packets). The reading is done cyclically, and a packet buffer per input is required to store the packet temporarily before it is read. The packets stored for the same output are read out on a first come first served basis. This reading is performed cyclically for the outputs. A packet buffer per output is needed to store the retrieved packet for transmission over a slot duration.

A memory location map is needed to manage the storage address of the packets. Also, each output maintains a FCFS list of the memory addresses of the packets destined for that output.

5.3 Fast Packet Switching

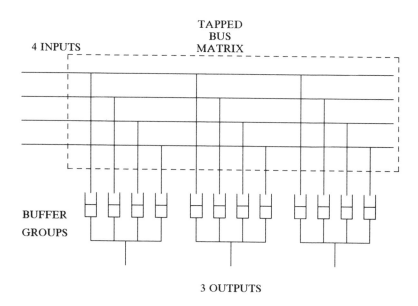

Figure 6. An Output Buffered Switch Node

The fast memory must be able to read or write at a speed proportional to the number of inputs or outputs. This speed requirement can be prohibitive when the switch node has a large number of inputs and outputs.

3. Multiple links, duplicated and appended switching networks

Besides the above methods which modify the functions of the switch nodes, we may instead change the topology of the switching network in ways shown in figure 8. Such changes produce a multiplicity of paths between each input and output.

The simplest change is to provide multiple links between nodes in the network, instead of a single link as shown in figure 5. Hence more than one packet from the inputs of the switch node can be delivered to the same switch node downstream. Each switch node now has to provide interconnections between more inputs and more outputs. The sequence of nodes traversed by a packet from an input to an output is

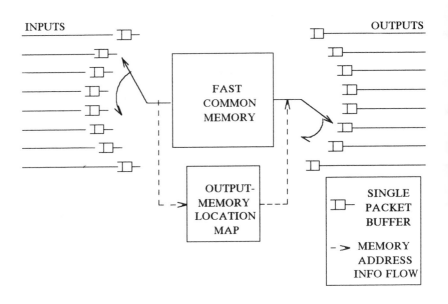

Figure 7. A Switching Node for a Multi-Stage Memory Switch

still unique, hence routing in this network is as simple as the single link case. The packet chooses one of the multiple outgoing links, or buffering within the switch node according to the local state of congestion.

The second method is to have several identical switching network laid out in parallel as shown in figure 8. We call this a duplicated switch. Each input is demultiplexed onto several switching planes. A packet may choose one of the switch planes randomly. The outputs of the switch planes are multiplexed onto the outputs of the switching network as shown. In effect, a duplicated switch reduces the load of each plane by a factor equal to the number of switch planes. Once the random choice is made, the packet follows a unique path in the switch plane chosen.

The third method is to cascade identical switching networks as shown in figure 8. We call this an appended switch. The cascade of two banyan networks is structurally similar to the Benes network

5.3 Fast Packet Switching 125

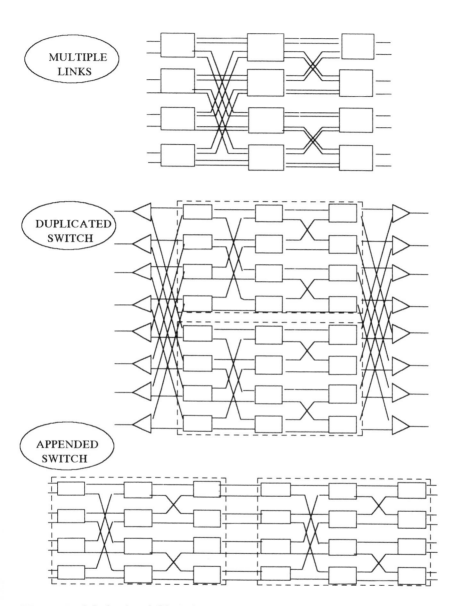

Figure 8. Methods of Changing the Topology of a Network

introduced in chapter 3. A packet at an input now has the liberty of choosing any output of the first half of the network. A packet in the middle of the cascaded network must then follow a unique path to the desired output. This freedom of routing in the first half of the network may alleviate congestion due to irregular traffic patterns, which shall be discussed in more details in section 5.5.

These networks with multiple paths between an input and an output create the problem of out-of-sequence packets. The information to be delivered is very often broken into packets to be delivered in a proper sequence. If different routes are used by these packets, the packets can arrive at the output not in the proper sequence, even though each node in the network has a first come first serve discipline. This out-of-sequence problem is due to the uneven degree of congestion in different routes, causing a varying degree of delay for the packets using these routes. Resequencing the packets is required when out-of-sequence of packets occurs.

5.4 Self-Routing Banyan Networks

A packet is self-routing if the header of the packet contains all the information needed to route through a switching network, and such routing is done using very simple mechanisms. We now describe a simple self-routing mechanism, which is shown in figure 9. Let the sequence of switch nodes in the route taken by a packet be $s_1 s_2 \cdots s_k \cdots s_K$. These nodes have $n_1, n_2, \cdots, n_k, \cdots, n_K$ outputs respectively. Instead of naming the nodes in a route, we can also name the output of the nodes in the route. Suppose the route uses the b_k-th output among the n_k outputs for node k. Then a self-routing address is given by the sequence $b_1 b_2 \cdots b_k \cdots b_K$. At switch node k, the packet is switched to the the b_k-th output. The address b_k is then discarded from the self-routing header. The packet continues to route through the remainder of the route using the residual self-routing header $b_{k+1} \cdots b_K$.

An interconnection network has the unique path property if the sequence of nodes connecting an input to an output is unique for all input-output pairs. One important class of multi-stage networks with the unique path property is the generic banyan network, for which some important examples are shown in figure 10. These networks include the banyan network introduced in chapter 4, and the baseline network introduced in chapter 3. Another important member of this

5.4 Self-Routing Banyan Networks

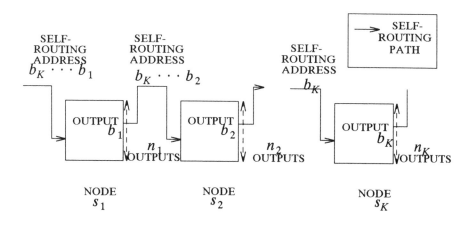

Figure 9. Self-Routing in Packet Switched Network

class is the shuffle exchange network, which we shall study in more detail. The flip network is simply the mirror image of the shuffle exchange network. All of these networks provide a unique path between every input-output pair. These networks are functionally and topologically equivalent in the sense that we can transform one network into the other by permuting switch nodes in each stage. We leave the discovery of the required permutation as an exercise for the reader.

Let us examine in particular the self-routing scheme for the shuffle exchange network. This scheme is illustrated in figure 11. The network has $N=2^n$ inputs and 2^n outputs, interconnected by n stages of 2^{n-1} binary switch nodes. Let us number the nodes in each stage top-down from 0 to $2^{n-1}-1$. Furthermore, we shall represent this numbering in binary $n-1$ bit format. Therefore, a node is specified by the stage it is in as well as its top-down numbering.

The links in a stage are numbered in a similar manner, namely in binary n bit format and top-down, as shown in figure 11 for the inputs and the outputs. Besides the inputs and the outputs to the network, the links are ordered top-down at the output of the switch nodes. For example, the outgoing links from the node numbered 01 at a specified stage are numbered 010 and 011 respectively. In other words, a 0 is

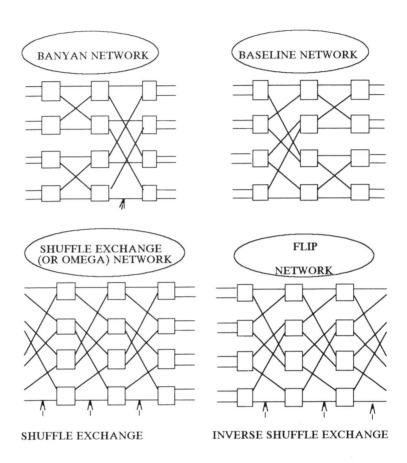

Figure 10. Some $N \log N$ Complexity Unique Route Networks

appended to the node number for the up outgoing link, and a 1 is appended to the node number for the down outgoing link.

A link connects two nodes in two consecutive stages, say stage k and $k+1$. The node at stage k has node numbering given by the link number without the right-most bit. The node at stage $k+1$ has node numbering given by the link number without the left-most bit. For example, the link 011 connects the node 01 to node 11 at the next stage.

Given this representation, the self-routing scheme for the shuffle exchange network is particularly simple. Suppose we have a packet at

5.4 Self-Routing Banyan Networks

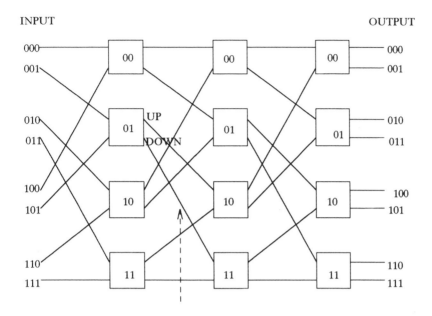

EDGE 011
CONNECTING NODE 01 (STRIKE OUT LSB OF EDGE) WITH
NODE 11 (STRIKE OUT MSB OF EDGE)

Figure 11. Representing Nodes and Links in a Self-Routing Shuffle Exchange Network

the input $a_1 a_2 \cdots a_n$. The packet is destined for the output $b_1 b_2 \cdots b_n$. We shall use the sequence $b_1 b_2 \cdots b_n$ as the self-routing address. We use the up link or down link at stage k according to whether $b_k = 0$ or $b_k = 1$. In this manner, the packet will end up at the desired output of the shuffle exchange network. More specifically, the packet visits the node $a_2 \cdots a_n$ in the first stage; uses the edge $a_2 \cdots a_n b_1$ to visit the node $a_3 \cdots a_n b_1$ at stage 2; ... ; uses the edge $a_{k+1} \cdots a_n b_1 \cdots b_k$ to visit the node $a_{k+2} \cdots a_n b_1 \cdots b_k$ at stage $k+1$; ... ; and finally arrives at the output edge $b_1 \cdots b_n$ from the node $b_1 \cdots b_{n-1}$ at stage n.

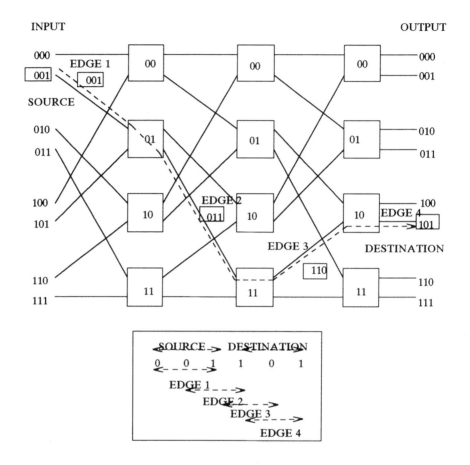

Figure 12. Self Routing in a Shuffle Exchange Network

We may also use this top-down numbering for nodes and links for the other networks shown in figure 10. We leave as an exercise for the reader to derive self-routing schemes for these networks.

This numbering also generates a simple proof for the following property which will be used in chapter 6. Consider in particular the shuffle exchange network. Suppose the self-routing addresses of packets at the inputs satisfy the following conditions. First, these addresses are strictly monotone in the sense that the destination

5.4 Self-Routing Banyan Networks

addresses are strictly increasing when considered top-down at the inputs. Second, these packets are compact in the sense that there is no idle input between any two inputs with packets. Then, the self-routing paths used by these packets do not share any link within the shuffle exchange network. Since no link is shared, it is not necessary to put buffering at the inputs of the internal nodes. Consequently, we can eliminate the internal buffers of the buffered banyan network if these monotone and compactness conditions are somehow achieved.

We proved in chapter 4 that the monotone and compactness conditions were sufficient for making the distribution network (a banyan network) non-blocking. The proof for the just stated property of the shuffle exchange network, though seemingly different from that given for the distribution network, actually uses similar insights. For this reason, we put the proof in the Appendix.

5.5 Combinatorial Limitations of Banyan Networks

A banyan network can realize $\exp_2(\frac{1}{2} N \log_2 N) = (N^N)^{1/2}$ input-output permutations as shown by the following argument. There are $\frac{1}{2} N \log_2 N$ switch nodes of size 2×2. Each 2×2 switch node can be in a cross state or a pass state (see figure 3 of chapter 4). Hence there are $\exp_2(\frac{1}{2} N \log_2 N)$ different states of the network. Each of these overall states realizes a distinct connection pattern, since the state defines N different link-disjoint paths, and each path uniquely defines an input-output connection by the unique path property of the banyan network. Hence, the number of realized permutation meets the upper bound of permutations allowed according to the node count.

However, the banyan network cannot realize all $N! \approx \exp_2(N \log_2 N)$ distinct input-output permutations. In fact, the banyan network has only half the node count according to the combinatorial upper bound. We have shown in chapter 3 that the Benes network, which comprises a baseline network and an inverse baseline network (both topologically equivalent to the banyan network), suffices for connecting all $N!$ permutations. Hence the fraction of permutations which can be realized by a banyan network goes to zero as $1/\sqrt{N!}$. Therefore, the banyan network is very blocking if used as a circuit switching network.

The combinatorial power of the banyan network can be increased significantly by the previously suggested methods of multiple links, duplicated switch or appended switch shown in figure 8. (The appended switch, which doubles the node count, is in effect a Benes network which can realize all permutations.) The exact number of permutations that can be realized is often fairly difficult to obtain.

Since most permutations cannot be realized by the banyan network, fast packet switching uses buffers at each internal link to ease local blocking. In the process, a random delay is introduced, which depends on the randomness of the packet addresses at the internal nodes of the banyan network. In the ideal case, we may assume that each packet at the inputs of the banyan network is independently and randomly addressed for the outputs. However, buffered banyan networks can be severely blocking if the destination addresses are not quite random. For example, certain outputs are more favored, thus creating what is termed "hot spots" at the outputs. Another problem arises when packets arriving at an input are more likely to go to the same output.

Consider for example that each input is repeatedly sending packets to a distinct output. The connection pattern is again a permutation, for which the banyan network can be very blocking. In the worst case, a link can be used by $\sqrt{N}$ input-output connections as shown in the following argument. An output of a switch node at stage $k \leq n/2$ of the banyan network can be reached by 2^k inputs to the network. At stage $k=n/2$, a node can be reached by $2^{n/2}=\sqrt{N}$ inputs, thus inducing the worst case number of connections using a link. However, this worst case number cannot be exceeded for a link beyond the $n/2$-th stage because the link can only reach less than $2^{n/2}$ outputs.

Assuming that all permutations are equally likely, the probability of this worst case offered load occurring is small for large N. We shall compute this probability in chapter 11. One means of tackling the extra load is by speeding up (relative to the input and output links) the transmission speed of the switching network. As shown in chapter 11, a speed up ranging from 4 to 8 may suffice. Besides speeding up, we may use multiple links in the network as shown in figure 8. Using multiple links has almost the same effect as a speed up, due to the equivalence between time multiplexing and space multiplexing. Alternatively, we may using duplicated switch planes, and send packets randomly into one of the switch planes.

More effectively, we may use an appended switch with random routing in the first half of the switching network as shown in figure 13. At each node in the first half of the appended switch, a packet is routed randomly to one of the two outputs. Consequently, a packet

5.5 Combinatorial Limitations of Banyan Networks

from an input is routed to a random link in the middle of the appended switch. Afterwards, the packet follows a unique path to the desired output in the second half of the network. This randomization of routing relieves link overloading resulting from address correlation of packets at an input.

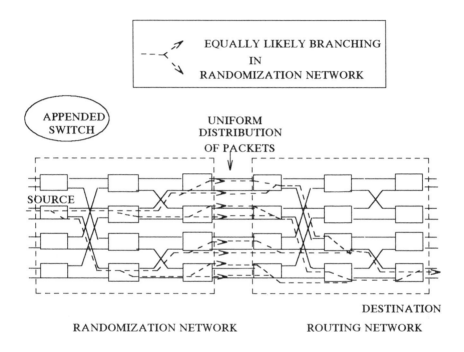

Figure 13. Randomized Routing for Internally Uneven Load

As mentioned before, the use of these methods may result in packets delivered out-of-sequence. Furthermore, the use of internal buffering generates an undesirable random delay. In the next chapter, we shall introduce alternative packet switching networks which do not use internal buffering, consequently eliminating some of these ill-effects. These networks use a sorting network to create a compact and monotone output pattern at the inputs of a banyan network. As a result, the banyan network becomes non-blocking, as proved in the following Appendix.

5.6 Appendix-Non-Blocking Conditions For Banyan Networks

Theorem:

Let $a = a_1 \cdots a_n$ and $a' = a'_1 \cdots a'_n$ be the input addresses of two packets for the shuffle exchange network. Let $b = b_1 \cdots b_n$ and $b' = b'_1 \cdots b'_n$ be the output addresses of these two packets. (The index 1 signifies the most significant bit.) With compact active inputs and monotone outputs, the self-routing paths of these two packets in the shuffle exchange network are link disjoint.

Proof:

Without loss of generality, we assume that $a' > a$ and $b' > b$. Since the active inputs are compact, the monotone output condition guarantees that the output address increases at least as fast as the input address, hence

$$b' - b \geq a' - a \tag{5.6.1}$$

On the other hand, consider the links traversed by these two packets from stage k nodes. As shown before, these links are represented by the part of the input-output addresses enclosed in square brackets:

$$a'_1 \cdots a'_k [\, a'_{k+1} \cdots a'_n \, b'_1 \cdots b'_k \,] \, b'_{k+1} \cdots b'_n$$

$$a_1 \cdots a_k [\, a_{k+1} \cdots a_n \, b_1 \cdots b_k \,] \, b_{k+1} \cdots b_n$$

If these two links are the same, then $a'_i = a_i$ for $k+1 \leq i \leq n$, and $b'_i = b_i$ for $1 \leq i \leq k$. Consider $a' - a$. Since a'_i must be different from a_i for at least one bit with index $1 \leq i \leq k$, we must have

$$a' - a \geq 2^{n-k} \tag{5.6.2}$$

On the other hand, $b' - b$ must be less than the value of the least significant bit in the square brackets, hence

$$2^{n-k} > b' - b \tag{5.6.3}$$

Equations (5.6.2) and (5.6.3) together imply

5.6 Appendix- Non-Blocking Conditions 135

$$a'-a > b'-b \quad (5.6.4)$$

which directly contradicts equation (5.6.1). Therefore, the two paths must be edge disjoint if the monotone and compactness conditions are satisfied. □

We shall use this theorem for internally non-blocking packet switches in the next chapter.

5.7 Exercises

1. Consider the three stage network with r middle stage switches. Hence there are r alternate paths. Suppose each of these paths is busy with independent probability q. We assume a random search algorithm for finding a free path. Show that the average number of search before a free path is found (or declared not found after searching all r paths) is given by $(1-q^r)/(1-q)$.

2. Consider the slot matching problem shown in figure 3 (and figure 7 in chapter 3). Assume the slots for the frames, each of size r slots, are occupied independently with probability p. Compute the probability that a connection requiring s slots can be made across a node, and comment on how this probability changes as a function of s and r.

3. Compare the relative merits in providing a buffer per input versus shared buffer at a switch node in terms of performance and implementation complexity.

4. Devise the self-routing scheme for the inverse banyan network and the baseline network shown in figure 10.

5. Show that the inverse banyan network and the baseline network are topologically equivalent to the shuffle exchange network. Specifically, how should the nodes in each stage of a banyan network or a baseline network network be permuted to obtain a shuffle exchange network? (Hint: start from the solution for the previous problem.)

6. Consider the hypercube network defined as follows. There are N nodes in the network, each named in binary from 0 to $N-1$. Each node i is connected to i' if the binary representations of i and i' differs only in 1 bit.

a. Characterize the complexity (number of nodes and links, fan-out, logical depth between any two nodes) of the network.

b. Devise a self-routing scheme between any two nodes in the hypercube network.

c. Assuming no buffering at each node, obtain an upper bound on the number of connection patterns which can be realized by the hypercube network. Compare that with the banyan network, which has more stages but a smaller fan-out.

d. Assuming buffering at each node, make a qualitative comparison for the performance of the hypercube network and the buffered banyan network.

7. Compare the three networks shown in figure 8, assuming input buffering at each node. We assume random routing at the 1×2 demultiplexers for the duplicated switch, and random routing as shown in figure 13 for the appended switch. Order the three networks for how well they deal with the following situations:

a. Unbalanced traffic pattern.

b. Minimizing mean delay in the network.

c. Minimizing the variance of delay in the network.

d. Minimizing the effort of reordering the packet sequence.

Give qualitative reasons for your ordering.

5.8 References

Banyan networks and their variants are widely used for interconnecting parallel processors. The tutorial by Wu *et. al.* contains a good collection of articles on the structure, use and performance of these networks. Variants of these networks were proposed by Lawrie (the Omega network), Stone (the shuffle exchange network), Goke *et. al.* (the banyan network), and several others. Batcher used the flip network for interconnecting memory locations in a distributed architecture call STARAN. These variants are known to be functionally equivalent, the proof of which can be found in the article by Wu *et. al.*

5.8 References

Turner first applied the internally buffered banyan network for integrated data and voice networking. This packet switching mode, called fast packet switching, is a significant departure from the earlier store and forward packet switching of data networking. Many improvements of buffered banyan networks have been proposed, including virtual cut-through proposed earlier by Kermani *et. al.*, load sharing using parallel fabric proposed by Lea, and cascaded and randomized routing discussed by DePrycker *et. al.* One proposal on using parallel buses and output queueing was given by Yeh *et. al.* The proof of the non-blocking condition in the Appendix follows the insight of Beckmann.

1. K. E. Batcher, "The flip network in STARAN," *Proc. 1976 Int. Conf. on Parallel Processing,* pp. 65-71, 1976.

2. W. H. Beckmann, "A contention-free routing scheme for a banyan network," AT&T Bell Laboratories Tech. Memo. TM 39351-529, 1982.

3. M. DePrycker and M. DeSomer, "Performance of a service independent switching network with distributed congtrol," *IEEE J. Select. Areas Commun.,* vol-5, pp. 1293-1301, Oct. 1987.

4. L. R. Goke and G. J. Lipovski, "Banyan networks for partitioning multiprocessing systems," *Proc. First Annual Computer Architecture Conf.,* Dec. 1973.

5. P. Kermani and L. Kleinrock, "Virtual cut-through: a new computer communication switching technique," *Computer Networks,* vol-3, pp. 167-286, 1979.

6. D. H. Lawrie, "Access and alignment of data in an array processor," *IEEE Trans. on Computers,* vol-24 no. 12, Dec. 1975.

7. C. T. Lea, "The load-sharing banyan network," *IEEE Trans. on Computers,* vol-35, pp. 1025-1034, Dec. 1986.

8. H. S. Stone, "Parallel processing with the perfect shuffle," *IEEE Trans. Computers,* vol.-29, no. 8, Feb. 1971.

9. J. S. Turner, "Design of an integrated services packet network," *IEEE Select Areas Commun.,* vol-4, Nov. 1986.

10. C. L. Wu and T. Y. Feng, *Tutorial: Interconnection Networks for Parallel and Distributed Processing,* IEEE Computer Society Press, 1984

11. C. L. Wu and T. Y. Feng, "Universality of the Shuffle-Exchange Network," *IEEE Trans. Computers,* vol-30, no. 5, May 1981.

12. Y. S. Yeh, M. G. Hluchyj and A. S. Acampora, "The Knockout switch: A simple, modular architecture for high-performance packet switching," *IEEE J. Selected Areas Commun.,* vol-5, no 8, Oct. 1987.

CHAPTER 6

Applying Sorting for Self-Routing and Non-Blocking Switches

- 6.1 Types of Blocking for a Packet Switch
- 6.2 Sorting Networks for Removing Internal Blocking
- 6.3 Resolving Output Conflicts for Packet Switching
- 6.4 Easing Head of Line Blocking
- 6.5 A Postlude - Integrated Circuit and Packet Switching
- 6.6 Appendix- Self-Routing Multi-Cast Switching

In Chapters 3 and 5, we discussed the complexity of path hunting and time slot matching algorithms for multi-stage circuit switching networks. While these algorithms have manageable complexity for circuit switching for telephony, they are not fast enough to allocate bandwidth for multi-rate and bursty services. To alleviate the processing bottleneck of bandwidth allocation, we introduced distributed algorithms using self-routing addresses and queueing at switch nodes for dealing with local congestion.

With the removal of the processing bottleneck, we have introduced randomness, which is often undesirable in communication systems. Consequently, we may experience excessive delay occasionally. In this chapter, we focus on internally non-blocking packet switches that reduce internal delay in the switching network. These switches are self-routing, and consequently allow distributed processing for routing packets.

6.1 Types of Blocking for Packet Switch

For packet switching, there are three types of blocking.

1. Internal blocking

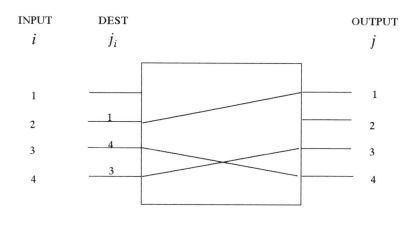

Figure 1. Internally Non-Blocking Switching

The first type of blocking occurs at the internal links of the switching network. Let us assume synchronous slot timing across the inputs of the switching network. An internally nonblocking switching network is one for which all distinct output connection requests at the inputs can be made within the slot time as shown in figure 1. Suppose the network has inputs i and outputs j, $1 \leq i, j \leq N$. During a slot time, the input i requests a connection to j_i. An internally non-

6.1 Types of Blocking for a Packet Switch

blocking switch can make all such connections provided that $j_i \neq j_{i'}$ if $i \neq i'$.

An $N \times N$ crossbar switch can be used for constructing an internally non-blocking packet switch. In this chapter, we shall focus on a new method of constructing a large $N \times N$ internally non-blocking and self-routing packet switching network via the mechanism of sorting.

2. Output blocking

The second kind of blocking occurs at the output of the switching network. An internally nonblocking switch can still block at the outputs due to conflicting requests, namely, the occurrence of $j_i = j_{i'}$ for some $i \neq i'$ as shown in figure 2. We call such an event an output conflict for output port j. When this event occurs, the switch should make a connection from at most one of the conflicting inputs to the requested output. The process of deciding which input is granted the right to connect to that output is called output conflict resolution.

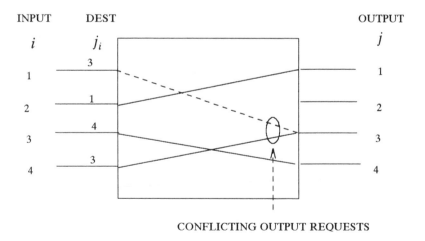

CONFLICTING OUTPUT REQUESTS

Figure 2. Output Blocking for Packet Switching

We see in figure 3 an example of output conflict for output 3, in which the request from input 1 conflicts with the request from input 4, and loses the contention for the output when the request from input 4 is granted.

A major distinction between a circuit switching node and a packet switching node is that a packet switching node must resolve output conflicts per slot, whereas for a circuit switching node, slots are scheduled for a frame with each input requesting connection to a distinct output for each slot. For a packet switch, an output conflict resolution mechanism is needed because the output slots are not assigned beforehand as in circuit switching.

When no prior schedule is made, a packet switch serves arriving packets at an input in a FCFS manner. Thus a packet which loses the contention may contend again in the next slot time. A random delay before a packet wins the contention is introduced.

When many such switching nodes with input queueing are used to build a multi-stage packet switching network with a large number of inputs and outputs, we experience more delay. By having a single internally non-blocking switch with a large number of inputs and outputs, queueing occurs at the inputs rather than internally.

An arbiter of outputs for the inputs can be used for resolving output conflicts for an $N \times N$ crossbar packet switch. In this chapter, we shall focus on using a sorting network for conflict resolution. This sorting network is the same as the one proposed for constructing internally non-blocking and self-routing packet switches.

3. Head of line (HOL) blocking

The third kind of blocking occurs at the input queues of the switching network. Consider two input queues with HOL packets contending for the same output. One of these HOL packet is granted the connection, while the other HOL packet is blocked. The blocked packet, due to the FCFS queueing discipline, may hinder the delivery of the next packet in the queue destined for a non-contended output. We illustrate this HOL blocking phenomenon in figure 3. Notice that no HOL packet seeks delivery to output 2, while a request for output 2 is blocked at input 1 because the HOL packet at input 1 is blocked by the HOL packet at input 4. HOL blocking reduces the throughput of a packet switch. The throughput and delay analysis for HOL blocking is given in chapter 10.

Having described these three forms of blocking, we now focus on how sorting can be used for dealing with them.

6.1 Types of Blocking for a Packet Switch

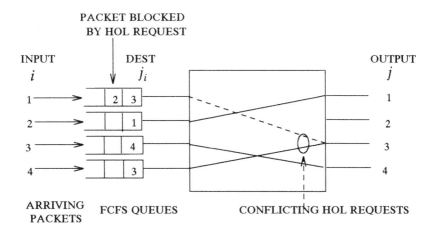

Figure 3. Head of Line Blocking for Packet Switching

Recall that in chapters 4 and 5, we showed that if the connection requests at the inputs of a banyan network are compact and in strictly increasing order, then the input-output paths are link disjoint, making the banyan network internally non-blocking. Consequently, one method of building an internally nonblocking network is to apply a sorting network in front of a banyan network to generate a strictly increasing order of destination addresses for the banyan network. This cascade of a sorting network and a banyan network is shown in figure 4, which illustrates the routes across this cascade for a set of packets with distinct destinations. We call this configuration the sort-banyan network.

A sorting network connects an input i, which has a destination request j_i, to an output of the sorting network according to the position of j_i in the sorted list of destination requests. This connection pattern for the sorting network is shown in figure 4. Sorting networks can be built by interconnecting nodes of smaller (such as 2×2) sorting networks as we shall show in the next section. Local comparisons of destination requests are made at these smaller sorting networks. Consequently, the packets are considered to be capable of self-routing through the network of sorting nodes. A packet can use its destination address as a self-routing address through the

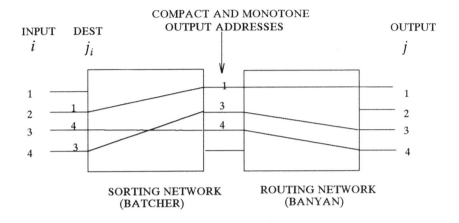

Figure 4. An Internally Non-Blocking and Self-Routing Switch

sort-banyan network.

As a refresher, the proof that compactness and monotonicity of destination addresses make a banyan network internally non-blocking can be grasped intuitively from figure 5. The banyan network is shown as a factorization of an $N \times N$ network into two stages of $N/2$ vertical 2×2 nodes and two horizontal $N/2 \times N/2$ nodes. The compactness and monotonicity of the destination addresses for the smaller $N/2 \times N/2$ nodes is evident in the recursive step.

We shall focus on building large sorting networks from 2×2 sorting elements in the next section. Afterwards we shall show how the sorting network can also be used for resolving output blocking and reducing HOL blocking.

6.2 Sorting Networks for Removing Internal Blocking

Sorting is intimately related to switching. If we have a permuted list of the vector $(1,2,3, \cdots ,N)$, we may use sorting to restore the original order. If we have a maximal connection pattern such that each input is connected to a distinct output, a sorting network for these destination requests would effectively perform the function of

6.2 Sorting Networks for Removing Internal Blocking　　　　　　145

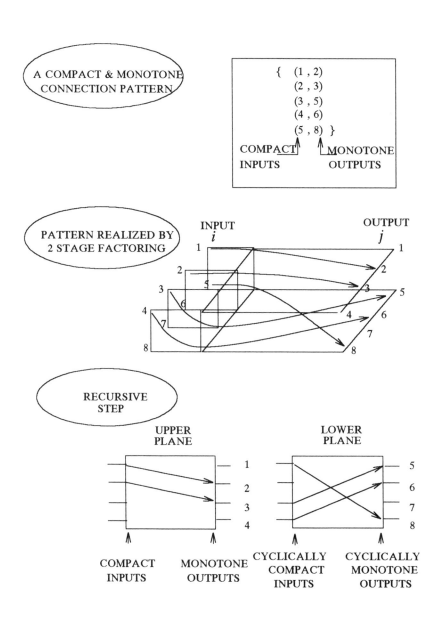

Figure 5. Non-Blocking Conditions for Banyan Network

6. Applying Sorting for Self-Routing and Non-Blocking Switches

switching. In other words, a switching network for a maximal connection pattern can be obtained from a sorting network by treating the 2×2 sorting elements as 2×2 switching elements.

We gave in chapter 4 an asymptotic lower bound of $N\log_2 N$ for the number of 2×2 switching nodes required to construct an $N \times N$ switching network. Consequently, the number of 2×2 sorting elements required to build a sorting network must also satisfy the same lower bound. Unfortunately, no sorting network found so far achieves this lower bound, or even comes within a small multiple of this lower bound.

Instead of using a sorting network that has a fixed interconnection pattern for its sorting elements, let us consider a more sequential sorting process. The following merge-sort algorithm satisfies the $N\log_2 N$ lower bound for the number of binary sorts.

Algorithm Merge-Sort

Input: An unsorted list $L_N = (l_1, \cdots, l_N)$

Procedure Sort:

$\text{Sort}(L_N) = \text{Merge}(\text{Sort}((l_1, \cdots, l_{\frac{1}{2}N})), \text{Sort}((l_{\frac{1}{2}N+1}, \cdots, l_N)))$

Procedure Merge:

$\text{Merge}((l_1, \cdots, l_m), (l'_1, \cdots, l'_{m'}))$

$= (l_1, \text{Merge}((l_2, \cdots, l_m), (l'_1, \cdots, l'_{m'})))$ if $l_1 \leq l'_1$

$= (l'_1, \text{Merge}((l_1, \cdots, l_m), (l'_2, \cdots, l'_{m'})))$ if $l_1 > l'_1$

Merge-Sort uses a divide and conquer strategy. The procedure Merge called by the procedure Sort takes two sorted lists and merges them by comparing the smallest elements in each of the two sorted lists as shown in figure 6.

Let us estimate the complexity of this sorting algorithm. The procedure of merging two sorted lists, each with $N/2$ numbers, requires N binary sorts. The total complexity of sorting N numbers is given by

$$C(N) = N + 2C(\frac{N}{2}) = N + 2(\frac{N}{2} + 2C(\frac{N}{4})) = \cdots = N\log_2 N$$

6.2 Sorting Networks for Removing Internal Blocking

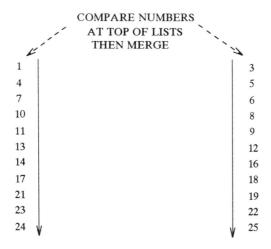

Figure 6. Merging Two Sorted List

Even though the total number of 2×2 sorts meets the $N\log_2 N$ lower bound required, this sorting algorithm takes at least $O(N)$ time due to the sequential nature of the procedure Merge. To reduce computation time, we parallelize the merge algorithm by using a network of 2×2 sorting elements. This algorithm was discovered by Batcher, based on the concept of bitonic sorting.

Definitions:

A bitonic list $L_N=(l_1,\cdots,l_k,\cdots,l_N)$ is a list such that for some $1 \leq k \leq N$, we have $l_1 \leq l_2 \leq \cdots \leq l_{k-1} \leq l_k$ and $l_k \geq l_{k+1} \geq \cdots \geq l_{N-1} \geq l_N$. In other words, the list monotonically increases to the k-th element of the list and then monotonically decreases.

A circular bitonic list is generated from a bitonic list as shown in figure 7 by joining the ends of the bitonic list and then breaking it into two equal halves.

A bitonic sorter is a sorting network that produces sorted outputs from bitonic inputs (or circular bitonic inputs). □

6. Applying Sorting for Self-Routing and Non-Blocking Switches

Bitonic lists have the following interesting property.

The Unique Cross Over Property:

When we compare a monotonically increasing list with a monotonically decreasing list, there is at most one position where the two lists cross over in their values as shown in figure 7. This unique cross over remains true for the two halves of the circular bitonic list, otherwise it would violate the order dictated by the solid and dotted arrows. □

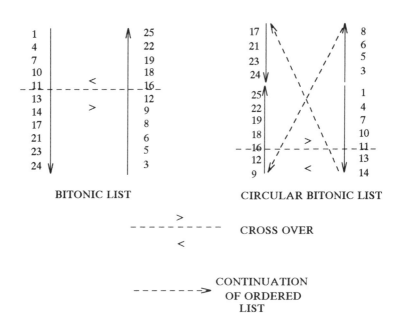

Figure 7. A Bitonic List, With Unique Cross Over

Using the unique cross over property, we prove the following theorem.

Batcher Theorem:

A circular bitonic list can be sorted by a banyan network for which the nodes are 2×2 sorters.

6.2 Sorting Networks for Removing Internal Blocking

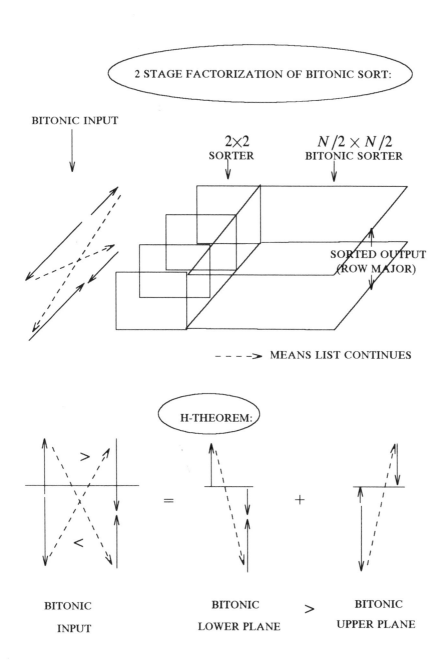

Figure 8. Recursive Construction of an $N \times N$ Bitonic Sorter

150   6. Applying Sorting for Self-Routing and Non-Blocking Switches

Proof:

Suppose we feed a circular bitonic list to the inputs of an $N \times N$ bitonic sorting network, factored into 2 stages of $N/2$ vertical 2×2 sorters and $\frac{1}{2}N \times \frac{1}{2}N$ horizontal bitonic sorters as shown in figure 8. This 2 stage factorization process generates the banyan network. We want to show that the outputs of the two planes of $\frac{1}{2}N \times \frac{1}{2}N$ bitonic sorters are sorted in a row major increasing order. To prove this fact, we have to show the following facts:

1. The inputs to each of the $\frac{1}{2}N \times \frac{1}{2}N$ bitonic sorters are circular bitonic lists.

2. The outputs of the upper bitonic sorter are each smaller than the outputs of the lower bitonic sorter.

To show these two facts, we have the following H-theorem which is a consequence of the unique cross over property. Let the horizontal line of the H mark the cross over point for the circular bitonic list represented by the vertical lines of the H. As we have shown, the position of this horizontal line is unique. After a side by side comparison by the 2×2 sorters, the larger values of the H are transferred to the lower $\frac{1}{2}N \times \frac{1}{2}N$ bitonic sorter, and the smaller values are transferred to the upper plane. Hence we have established fact 2.

To show fact 1, we observe how the H is divided as shown in figure 8. We see readily each part of the divided H consists of 2 sorted lists, which would appear as a circular bitonic list to each of the $\frac{1}{2}N \times \frac{1}{2}N$ bitonic sorter.   □

An alternative proof for bitonic sorting can be found in the exercises, using a well-known property of sorting networks called the 0-1 principle and a sorting algorithm called shear sort.

We can now construct a Batcher sorting network from a recursive application of bitonic sorting as shown in figure 9. Using 2 2×2 sorting elements, we form bitonic lists of 4 elements, which are then sorted by a 4×4 bitonic sorting network. We then take 2 sorted lists of 4 elements each and sort them by an 8×8 bitonic sort network; etc. The resulting sorting network is called a Batcher sorting network, which consists of multi-stages of banyan sorting networks of increasing size. The successive stages of bitonic sorters are connected by links with a shuffle exchange pattern as shown in figure 9. (The shuffle exchange is needed when the network is shown by a two dimensional representation. The reader may arrive at this conclusion through a

6.2 Sorting Networks for Removing Internal Blocking

careful comparison of a bitonic sorter in figure 9 with its three dimensional representation in figure 8.)

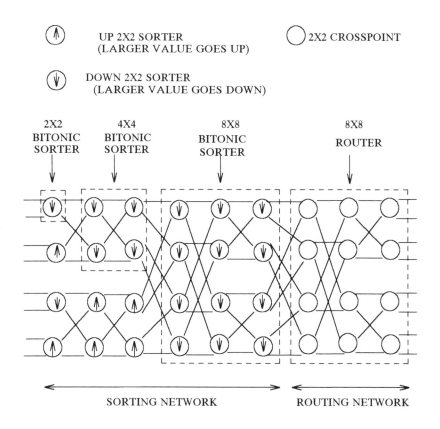

Figure 9. The Batcher Sorting Network

Let us estimate the total number of 2×2 sorting elements needed for the Batcher sorting network. In order to sort 2^n elements, we need n stages of bitonic sorting. The stage i bitonic sorter has i stages of 2×2 sorting elements itself. Hence the total number of stages of the sorting network is $\frac{1}{2} n(n+1)$. Each stage has $\frac{1}{2}N$ binary sorters. The total number of binary sorters is then $0.25\ N\ \log_2 N\ (\log_2 N + 1)$.

This number is quite a bit larger, particularly for large N, than the $N\log_2 N$ lower bound. There are other sorting networks of complexity $CN\log_2 N$ in which the constant C is quite large, which makes such constructions impractical. The Batcher algorithm is the least complex among all known sorting networks for most practical values of N.

6.3 Resolving Output Conflicts for Packet Switching

As noted before, a packet switch does not maintain a schedule for dedicating time slots in a frame to the inputs. Therefore an output conflict resolution process is needed on a slot by slot basis.

One method of resolving conflict is to have each output poll the inputs in a round robin fashion. The first polled input with a request for that output wins the contention. Polling can be implemented in many ways. One way is through the use of a token ring which circulates output tokens around the inputs. A token generator issues a token for each output for circulation in the ring. The tokens for the outputs are issued sequentially. A token is seized when it arrives at the first input contending for the output associated with that token. The input with the output token is allowed to make a connection to that output, such as through a sort-banyan routing network, or through an $N \times N$ crossbar switching network.

Output conflict resolution by polling has the following drawbacks:

1. The amount of polling per output grows linearly with the number of inputs. This limits the number of inputs if we have to poll all inputs within a fixed slot time.

2. The inputs which are polled earlier have an unfair advantage in seizing the time slot.

We need an alternative mechanism for fair and fast resolution of output conflicts for packet switches with a large number of inputs.

Fortunately, the sorting network which makes the banyan network internally non-blocking can also be used to resolve output conflicts as shown in figure 10. When the self-routing addresses are sorted, duplicated output requests would appear adjacent to each other in the sorted order.

Suppose the self-routing address appearing at the output k of the sorting network is the same as that at output $k-1$. Then we can locally decide that the packet appearing on output k loses the

6.3 Resolving Output Conflicts for Packet Switching

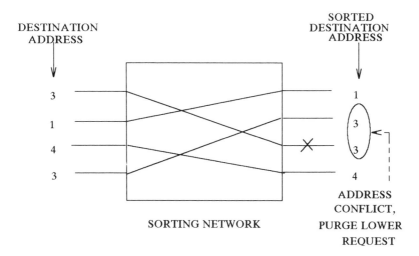

Figure 10. The Using of Sorting for Detecting Output Contention

contention. If the two addresses are different, then the packet on output k can be delivered to the desired output.

The sorting network can also easily discern packet priority in this process of purging output conflicts. All we have to do is to append a priority field to the self-routing address. The sorting network would sort the appended addresses in such a way that the higher priority packets are placed at a favorable position before purging. Hence the packet with the highest priority wins the adjacent line conflict resolution process. As we shall see later, this built-in priority feature is very important for integrating circuit switching and packet switching in a sort-banyan network.

However, purging duplicated requests in the sorted order can cause blocking in the banyan network if we allow the surviving packets to route through the banyan network, since the input requests to the banyan network may not be compact any more. Hence we need some structural modification to the sort-banyan network to achieve the function of packet switching. There are many approaches, but we shall cite only 2 for illustration.

1. Sort-purge-concentrate with re-entry for blocked packets

First, we may remove the gaps in the purged requests by concentrating the output requests of the sorting network as shown in figure 11. Obviously, such concentration can be performed by another sorting network, with the destination address of the purged packets changed to a large value so that the purged packets would be sorted to the bottom of the sorting network. Alternatively, we may use a compact superconcentrator discussed in chapter 4, which can be made self-routing as shall be shown in the Appendix. The concentrated requests are then routed through the banyan network.

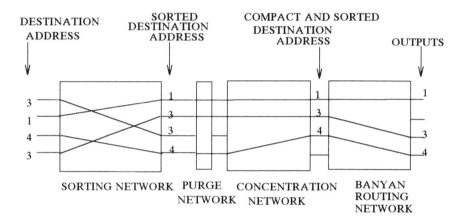

Figure 11. Sort-Purge-Concentrate for Packeting Delivery

There are two ways of dealing with purged packets. The first way requires each output which has received a packet to acknowledge the input which sent the packet. The acknowledgement packet can be sent through the same network of figure 11 in the following manner. By pairing the output j of the network with the input j of the network, a delivered packet with the source-destination address (i,j) now sends an acknowledgement packet with destination address i from input j of the network. The destination addresses of the acknowledgement packets are distinct, thereby guaranteeing their successful delivery. We require each input to store the packet in an input buffer until the packet is acknowledged for the successful delivery. The inputs not

6.3 Resolving Output Conflicts for Packet Switching

acknowledged will try to send the same packet again in the next time slot.

Alternatively, we may use a re-entry scheme for the purged packets as shown in figure 12. The purged packets are concentrated by the second sorting network after their addresses are changed to a very large value. The purged packets are then buffered for re-entry to the front end of the switching network in the next slot time.

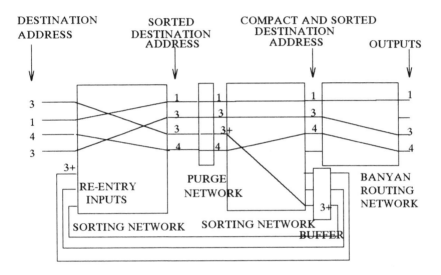

3+   ALTERED ADDRESS OF PURGED PACKET

Figure 12. Re-Entry of Purged Packets for Retrial

One drawback of the re-entry approach is that extra input ports have to be dedicated for re-entry. Also, care must be exercised to avoid packets delivered out-of-sequence, or packets lost due to an excessive number of re-entering packets.

2. Probe-acknowledge-send with input buffering for blocked packets

6. Applying Sorting for Self-Routing and Non-Blocking Switches

Instead of sending a packet directly, we may have each input with a HOL packet probe the availability of the desired output before sending a packet. We call this probing process the request phase. We then need a second phase for acknowledging each input from the output of the sorting network about the result of the conflict resolution. We call this the acknowledgment phase. Afterwards, the acknowledged input sends its packet in the third phase, the packet transmission phase. To illustrate this three-phase procedure, we discuss the details of two implementations.

For the first implementation, each input with a HOL packet sends a request packet through the sorting network. The request packet consists of simply the destination address of the HOL packet. At the output of the sorting network, output conflicts among requests are resolved by purging. To acknowledge the successful inputs, we propagate the acknowledgement backwards through the same path used by the request packet as shown in figure 13. The acknowledged input would then send the HOL packet through the sorting network. Since all except one duplicated requests are purged, we no longer have output conflict. Hence the sort-banyan network is sufficient for non-blocking delivery of packets from the acknowledged inputs. An input receiving no acknowledgement stores the HOL packet in an input buffer, and contends for the output again in the next time slot.

Instead of back propagating the acknowledgement signal through the sorting network, we may send the acknowledgement by the sort-banyan network in a forward manner, using the three-phase algorithm illustrated in figure 14. Suppose at the output k of the sorting network, the self-routing address is different from that at the output $k-1$. Hence we have to send an acknowledgement from output k of the sorting network to the requesting input.

By bringing a fixed connection from the output k of the sorting network to the input k of the sorting network, we may send the acknowledgement from the input k to output i of the sort-banyan network. Therefore, the request packet sent in the first phase must contain not only the destination address j_i, but also the source address i which is used as a self-routing header for the acknowledgement. Since the inputs i winning the contention are distinct, the acknowledgements are guaranteed non-blocking delivery by the sort-banyan network in the second phase. The third phase for delivering packets from acknowledged inputs and input buffering for not acknowledged inputs are the same as that for the back propagation scheme.

6.3 Resolving Output Conflicts for Packet Switching

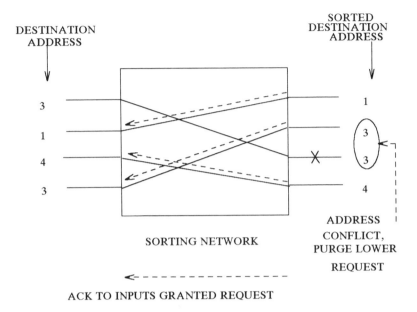

Figure 13. Back Propagation of Acknowledgement

The extra phases of request and acknowledgement require an overhead to the utilization of the switching network. This overhead is dominated by the passage time of the request and acknowledgement packets through the sorting network. Hence the overhead is roughly given by $(\log_2 N)^2$ bit time (twice the number of stages of a Batcher sorting network). The switching network has to be speeded up in order to accommodate the extra phases. There are many clever methods to reduce this overhead. These methods take advantage of the fact that the length of the request packet or the acknowledgement packet is relatively short compared to the number of stages of the Batcher sorting network, hence most of the overhead results from the passage time of the these packets through the Batcher network. Hence staggering the phases of various time slots gives a more pipelined three-phase algorithm. We leave as an exercise for the reader to derive the proper way to stagger the phases.

6. Applying Sorting for Self-Routing and Non-Blocking Switches

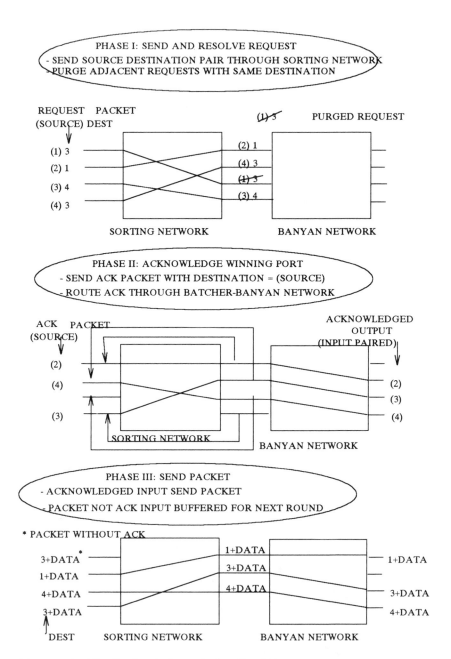

Figure 14. The Probe-Acknowledge-Send Process

6.4 Easing Head of Line Blocking

Essentially, there are two basic approaches for reducing HOL blocking.

1. Allow packets behind the HOL packet to contend for outputs

Take the token ring method described at the beginning of section 6.3 as an example. When an output token arrives at an input, we may allow the HOL packet as well as those behind to seize the token. Once a token is seized by any packet in the input queue, no other output tokens may be seized by that input because an input can send at most one packet per time slot.

2. Allow multiple delivery of conflicting HOL packets to an output buffer

Take the output buffering scheme shown in figure 6 of chapter 5 as an example. The tapped bus matrix provides many paths, hence enabling all conflicting HOL packets to be delivered to the output within a slot time.

These two examples become impractical for large number of inputs and outputs. Alternatively, these two approaches can also be used for the sort-banyan network.

1. Multiple rounds of arbitration for the sort-banyan network

The sorting network can also be used in multiple rounds of request-acknowledgement for reducing the HOL blocking as shown in figure 15. After 1 round of request-acknowledgement, those HOL packets which lose the conflict resolution allow the packet behind the HOL packet to contend for a second round of request-acknowledgement. These second attempts must not interfere with the requests granted during the first round. This is achieved by allowing the HOL packets which won the first round to contend again in the second round with higher priority than those packets behind the HOL packets. This guarantees that the first round winners would win again in the second round. The same procedure may be repeated for a third round for the third packet in the queue. The throughput improvement versus the number of rounds of arbitration is evaluated in the exercises of chapter 10. However, the extra throughput achieved by reducing the HOL blocking requires an overhead use of the switching network for request-acknowledgement. This overhead is proportional to the number of rounds. Again, clever means can be used to reduce

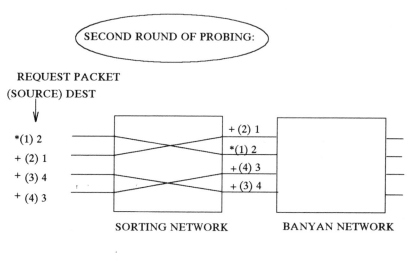

Figure 15. Multiple Rounds of Arbitration to Remove HOL Blocking

the overhead.

2. Multiple planes of sort-banyan networks

By speeding up the switching network with respect to the transmission speed of the inputs and outputs, input queueing effects are reduced. In the process, we have shifted the queueing effect from the input queues to the output queues. Instead of an actual speed up, the input queues can be served faster by the use of parallel switch networks as shown in figure 16.

Each switch network operates the three-phase algorithm, with the second switch network staggered by half a time slot behind the first

6.4 Easing Head of Line Blocking

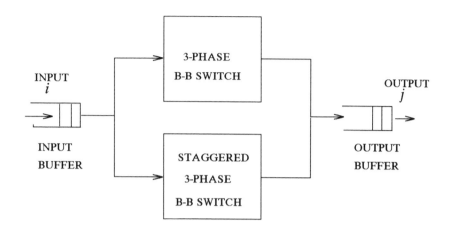

Figure 16. Use of Parallel Switching Planes for Speed-Up

switch network. The HOL packets contend for the first switch network. Since the first two phases of the three-phase algorithm finish in less than half a slot time, packets may move up the queue if the HOL packet is served. The resulting HOL packets contend for the second switch network. This process alternates between the two switch networks. More generally, we may allow more than two switch networks to operate in parallel, provided that adjacent switch networks are staggered by a period of time longer than that required by the first two phases of the three-phase algorithm.

6.5 A Postlude - Integrated Circuit and Packet Switching

To conclude the first half of this book on switching theory, we take an integrated view of the various kinds of switching discussed. We started with circuit switching, moved onto multi-point switching, then self-routing packet switching, and finally internally non-blocking packet switching. Ideally, we would like to provide all such switching in a single switch fabric.

Of particular interests is the capability to integrate circuit switching and packet switching. Recall that circuit switching dedicates time slots of a frame for each connection, whereas packet switching allows contention for time slots on a slot by slot basis. While circuit switching guarantees zero delay variance, packet switching facilitates more efficient and dynamic allocation of bandwidth resources. Ideally, an integrated services network should be capable of emulating circuit switching for fixed bit rate and delay sensitive services, and providing packet switching for services with fluctuating bit rates. Consequently, a switching system should be flexible in trading off efficiency of bandwidth allocation versus control of delay.

How should this tradeoff be placed? Consider services for which packets are generated periodically. The delivery of these packets at the receive end is not quite periodic due to queueing in the network. This disturbance depends on the load of the network. Let us define the packet jitter as the standard deviation of the packet arrival time, divided by the expected packet interarrival time. It follows then packet jitter is strongly influenced by the bit rate of the service, which is inversely proportional to the packet interarrival time. In other words, a high bit rate service would suffer a higher degree of packet jitter than a low bit rate service. Furthermore, packets of a high bit rate service tend to cluster together after they move through several packet switches in tandem. This snowball effect can seriously affect switching delay, throughput, and buffer overflow probability.

Therefore, high bit rate connections should be circuit switched. Low bit rate connections may be packet switched on a FCFS basis. In an integrated circuit and packet switching network, priority classes should be assigned to distinguish a circuit connection from a packet connection. We allow packet switched connections to seize a time slot if the circuit switched connection do not utilize its assigned time slot in a frame. This integrated multiplexing scheme (figure 7 of chapter 2) was termed hybrid TDM in section 2.3.

The sort-banyan network, with its non-blocking, self-routing, and built-in priority functions, provides an excellent mean to integrate circuit and packet switching. We shall use the sort-banyan network to illustrate an implementation of an integrated switching system for interconnecting hybrid TDM links.

Using sorting networks, it is possible to build self-routing and non-blocking switches with a large number (N) of inputs and outputs. Output conflicts are resolved by sorting for packet switching, and avoided for circuit switching by assigning time slots in a frame. For circuit switched connections, the time multiplexed sort-banyan switch

6.5 A Postlude - Integrated Circuit and Packet Switching

is a two-stage TS circuit switched network. When a circuit switched slot is not used by a circuit switched connection, the slot may be used by packet switched connections with a lower priority.

Slot assignment for circuit switching can be processed in a distributed manner, using the same sort-banyan network for the communications required by the distributed process. Let L be the number of time slots in a frame. Let each input i maintains an L-vector $\underline{u}_i$ indicating the assigned input time slots. Similarly, let each output j maintains an L-vector $\underline{v}_j$ indicating the assigned output time slots. Suppose we want to assign a number c of time slots for a connection between input i and output j. The connection can be established if there are at least c time slots commonly vacant to both $\underline{u}_i$ and $\underline{v}_j$. At call setup, the input i sends to j a packet containing the information $\underline{u}_i$ and c. This packet is routed through the sort-banyan packet network in a packet switched mode. The output j then checks for common vacant time slots in $\underline{u}_i$ and $\underline{v}_j$. If the c requested slots can be assigned, the output j sends to i a packet marking the locations of the assigned time slots. The input i and output j subsequently updates the slot assigned vectors $\underline{u}_i$ and $\underline{v}_j$ respectively.

This 2 stage TS network is rearrangeably non-blocking. To make it strict-sense non-blocking, we have to speed up the sort-banyan network by a factor of two with respect to the speed of the inputs and outputs. Alternatively, we may have staggered parallel switch networks described in the previous section for strict-sense non-blocking circuit switching. When we integrate packet and circuit switched traffic, more speed up is required so that packet switched connections may not suffer from excessive queueing delay in the background of higher priority circuit switched traffic.

After integrating circuit and packet switched connections, the next step of integration is to provide multi-point connections (both circuit switched and packet switched) in addition to point-to-point connections. Much less is known about this broad theme of integration. Most self-routing multi-point networks to date factor the function of multi-casting into the functions of copy and point-to-point routing. Given an integrated circuit and packet realization for point-to-point routing, we need a self-routing copy network to add packet switched multi-cast functions to the network. The circuit switched copy network of chapter 4 can be converted into a self-routing packet switch. We leave the description of the self-routing scheme to the Appendix. However, the problem of integrating circuit switched and packet switched multi-cast traffic is not yet well understood.

6.6 Appendix- Self-Routing Multi-Point Switching

The major difficulty to make a multi-point network self-routing is that instead of a single destination address, a multi-point packet has a number of destination addresses. If each packet carries all such addresses, the packet header would have variable length and could be lengthy. To avoid this header problem, we may labeled the packet to be copied by a virtual address called a Broadcast Channel Number (BCN) as well as by the number of copies to be made. After the copies are made, we give each copy a destination address. Therefore, a directory is needed to give each copy with the same BCN a distinct destination address. We are going to describe the self-routing multi-point switch proposed by Lee.

We showed in chapter 4 how a multi-point circuit switch was constructed using the cascade of a copy network and a point-to-point network. The copy network was constructed using the cascade of a compact superconcentrator and a copy distribution network. We may use the reversed banyan network for the compact superconcentrator, and the multi-cast banyan network for the copy distribution network.

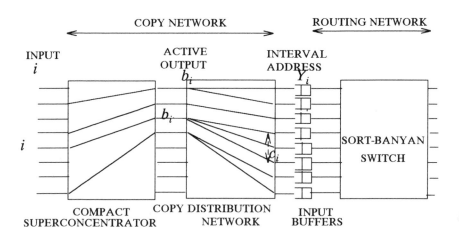

Figure 17. A Multi-Point Packet Switch

Since the sort-banyan network is self-routing, it remains to be shown that the copy network can be made self-routing.

6.6 Appendix- Self-Routing Multi-Point Switching

For each slot time, let $c_i \geq 0$ be the number of copies to be made for the packet at input i of the copy network, $0 \leq i \leq N-1$. Let the activity bit $a_i = 0$ if $c_i = 0$ and $a_i = 1$ if $c_i \geq 1$. The function of the compact superconcentrator is to connect an active input i with $a_i = 1$ to the output $b_i = \sum_{k=0}^{i-1} a_k$.

The function of the copy distribution network is to establish a multi-cast tree from its input b_i to the outputs from $m_i = \sum_{k=0}^{i-1} c_k$ to $M_i = \sum_{k=0}^{i} c_k - 1$, as shown in figure 17. Obviously the interval $[m_i, M_i]$, bounded by the minimum and maximum of the outputs in the interval, has c_i outputs. The reason for assigning this output interval was given in chapter 4.

To make the copy network self-routing, we have to provide the self-routing addresses b_i for the compact superconcentrator, and the self-routing interval address $Y_i = [m_i, M_i]$ for the copy distribution network. These two running sums m_i and M_i, together with the running sum b_i can be found by using a running sum adder network shown in figure 18.

Notice that an adder element in the network adds the number at the top and the number at the left of the adder and passes the sum to the right. The adder also passes the number from the top to the bottom unchanged. The adders are placed in the network in the following manner. Consider the i-th horizontal line in the adder network, with c_i as input at the left and M_i as output on the right. Let us represent i by its binary $\log_2 N$ bits representation, with the most significant bit at the right. The adders are placed on the line according to this binary representation. Take $i=3$ with the binary representation 110 as an example. The adder placement is 110 as shown in figure 18. The resulting output of the line is indeed M_i as defined.

The address b_i can be used as a self-routing address for the compact superconcentrator (implemented by a reversed banyan network) in a manner similar to that for fast packet switching (implemented by a banyan network). The interval address Y_i represented by its two bounds can be used for self-routing in the broadcast banyan network by the following interval splitting algorithm, which is illustrated in figure 19. The corresponding multi-cast tree in

166   6. Applying Sorting for Self-Routing and Non-Blocking Switches

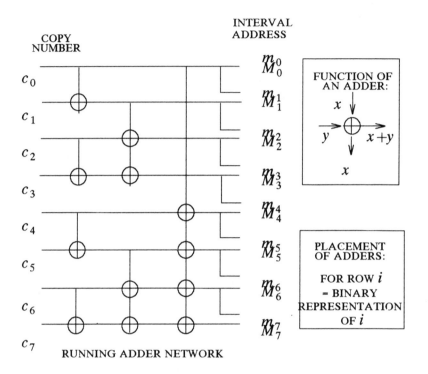

Figure 18. Running Sum Adder Network

the banyan copy network is illustrated in figure 20. This tree originates from the input $i=0100$ and terminates on the interval $Y_i=[0101,1010]$.

The n-bit binary numbers in a rectangle in figure 19 are the outputs in an interval. The two binary numbers above the rectangle are the minimum and the maximum of the numbers in the rectangle. At stage k of the broadcast banyan network, we look at the k-th bit, counted from the left and indicated by the arrow in figure 19, for all numbers in a rectangle. Three cases can occur:

1. The k-th bits are all 0's -

6.6 Appendix- Self-Routing Multi-Point Switching

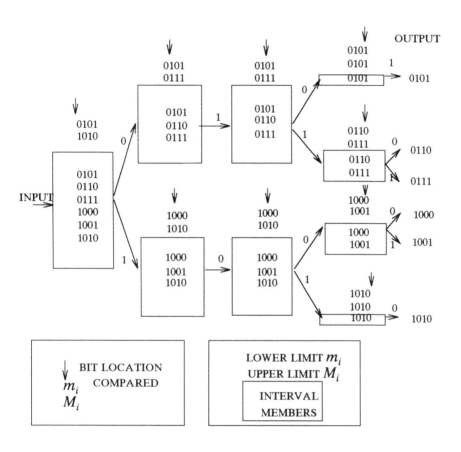

Figure 19. The Interval Splitting Algorithm

- Do not split the interval. Pass the interval from the current node in stage k to the next node in stage $k+1$ by the UPPER outgoing link. The interval is coded by the same min and max of the original interval.
- Extend the multi-cast tree using the UPPER outgoing link.

2. The k-th bits are all 1's -

- Do not split the interval. Pass the interval from the current node in stage k to the next node in stage $k+1$ by the LOWER outgoing link. The interval is coded by the same min and max of the original interval.
- Extend the multi-cast tree using the LOWER outgoing link.

3. Some k-th bits are 0's and some are 1's -
    - Split the interval into two sets according to whether the k-th bit is 0 or 1.
    - Pass the interval associated with 0 from the current node in stage k to the next node in stage $k+1$ by the UPPER outgoing link. This new interval has the same min as the original interval. The max of the new interval is given by replacing the $k+1$ to n-th bits of min by 1's.
    - Pass the interval associated with 1 from the current node in stage k to the next node in stage $k+1$ by the LOWER outgoing link. This new interval has the same max as the original interval. The min of the new interval is given by replacing the $k+1$ to the n-th bit of max by 0's.
    - Extend the multi-cast tree using both the UPPER and LOWER outgoing links. Thus two copies of a packet are made.

Notice at stage k, the splitting decision can be made just by comparing the k-th bit of min and max. If the k-th bits of min and max are different, the new min and max for the new intervals can be generated using very simple logic.

After the copies are made by the copy network, a point-to-point address has to be given to each copy. Before describing the mechanism for obtaining the address, we describe a method to give each copy a unique copy index e, such that $0 \leq e \leq c_i - 1$.

We allow the packets to remember the original m_i as they make copies of themselves through the copy network. Suppose we have a copy arriving at the output $y_i \in Y_i = [m_i, M_i]$ of the copy network. We can then obtain the copy index $e = y_i - m_i$.

Given the broadcast channel number BCN and the copy index e, we may obtain a destination address from an address translation table at each input of the point-to-point packet switch. The address translation table contains a list of destinations for each BCN. The

6.6 Appendix- Self-Routing Multi-Point Switching

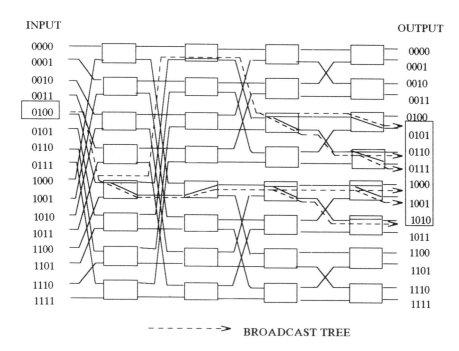

Figure 20. A Multi-Cast Tree in a Banyan Network

table is updated during call setup, which assigns a BCN to the call.

To summarize the above functions, we describe the format of the packet header shown in figure 21. The first field contains a self-routing address through the compact superconcentrator. The second field contains a bit interleaved pair of self-routing addresses m_i and M_i through the banyan copy network. The pair is modified through the network according to the interval splitting algorithm. The third field contains the original m_i for computing e at the output of the copy network. Given e and the fourth field which contains the BCN, the destination address of the copy can be obtained from an address translation table.

For a given BCN, the interval boundaries m_i and M_i change from slot to slot because they depend on how many copies are requested by the other BCNs. Consequently, copies with the same e and BCN may appear at different inputs of the point-to-point packet switch. Hence it is necessary that the mapping of e and BCN onto a destination

INFORMATION	BCN (BROADCAST CHANNEL NUMBER)	m_i (INDEX REFERENCE)	m_i, M_i (MIN,MAX BIT-INTERLEAVED)	b_i (ADDRESS OF CONCENTRATOR)

Figure 21. Packet Header Format for the Copy Network

address be available at all inputs. Also, packets can arrive out-of-sequence because they may use different inputs of the point-to-point packet switch with different degree of congestion.

A packet copy network also has the problem of copy request overflow. Let M be the number of copies that can be made by the copy network per slot. Copy request overflow occurs when there is some input i such that $M_i > M$ and $M_{i-1} \leq M$. Therefore for all $i' \geq i$, the copy requests cannot be satisfied. In this case, the value $M_{i'} > M$ computed by running sum adder network also serves to inform the input i' to hold the copy request until the next time slot. It is important to size the copy network properly so that the probability of copy request overflow is acceptably small.

6.7 Exercises

1. The 0-1 principle for sorting networks states that if a network of sorting elements can sort all patterns of inputs for which each input is either a 0 or a 1, then it can sort all patterns of inputs without the 0-1 restriction. To prove this statement, consider the converse: If a network cannot sort the inputs without the 0-1 restriction, then it cannot sort the inputs with the 0-1 restriction either.

    a. A monotonic function $f(x)$ satisfies $f(x) \leq f(y)$ whenever $x \leq y$. Show that if a sorting network produces the outputs $(y_1, \cdots, y_n)$ from the the inputs $(x_1, \cdots, x_n)$, then the network will produce the output $(f(y_1), \cdots, f(y_n))$ from

6.7 Exercises

the inputs $(f(x_1), \cdots, f(x_n))$.

    b. Prove the converse statement. (Hint: Consider f being a binary function which maps all numbers less than y_i into 0 and all numbers greater than or equal to y_i into 1, for some $\dot y_i$ which is not sorted, in other words, $y_i > y_{i+1}$.)

2. Let us compare the complexity of sorting zeros and ones versus sorting a permutation of 0 to $N-1$.

    a. How many different 0-1 input patterns are there? Compare that with the number $N!$ of permutations of N objects.

    b. How would you sort a 0-1 list sequentially, with a time complexity meeting the lower bound derived in part a?

    c. What can you say about sorting networks, in light of the 0-1 principle and the results of parts a and b?

3. We describe a shear sort algorithm. Consider sorting $N = p \times q$ numbers, which we assume to be 0 or 1 using the 0-1 principle. Let us put these N numbers into a square array with p rows and q columns. The shear sort algorithm works as follows:

    1. Sort each even row in increasing order from left to right. Sort each odd row in increasing order from right to left.

    2. Sort each column in increasing order from top to bottom.

    3. If each row, with possibly one row as exception, contains all 0 or all 1, sort all rows one more time in increasing order from left to right. Stop. The array is now sorted in a row major fashion from left to right and from top to bottom. Otherwise, go to step 1.

Let us now analyze this algorithm

    a. To get acquainted with the algorithm, randomly fill an 8×8 array with 0 and 1, and then apply the shear sort algorithm to the array. List the array generated by each step. What is happening in each step? How many times does the algorithm visit step 1?

    b. Prove that the algorithm visits step 1 at most $\log_2 p$ times. (Hint: Prove that the number of "dirty rows" containing both 0 and 1 is reduced by a factor of 2 by applying step 1 and 2.)

4. A bitonic sorter is a special case for the shear sort algorithm, with $p=2$. Use the proof for shear sort to show that a bitonic sorter generates a sorted list.

5. Use the shear sort algorithm to construct a sorting network containing $p \times p$ and $q \times q$ sorters. Also, try to lay out these $p \times p$ sorters in parallel and orthogonal to the $q \times q$ sorters, so that the sorting network assumes a 3-dimensional form, similar to the way we represented switching network in chapters 3 and 4.

6. Derive a 3 dimensional form for the $2^n \times 2^n$ Batcher sorting network. (Hint: Look at the sorting network in figure 9. Treat the front end bitonic sorters of size $2 \times 2, 4 \times 4, \cdots, 2^{n/2} \times 2^{n/2}$ as $2^{n/2}$ parallel $2^{n/2} \times 2^{n/2}$ sorters, and treat each $2^i \times 2^i$ bitonic sorter for $n/2 < i \leq n$ separately.) How many stages of orthogonal planes are there? Compare with that for the shear sorting network.

7. Carefully account for the total time required for the three phase algorithm. Evaluate the overhead required for $N=1024$ inputs and a packet size of 1024 bits.

8. Devise means of staggering the phases of the three phase algorithm in various time slots for minimizing the overhead required. How much overhead is saved, and at what cost in complexity and delay?

9. The three phase algorithm can be modified to deliver $k>1$ packets to an output per slot. Consequently, queueing is shifted from the inputs to the outputs.

    a. Devise a mean to purge all but k requests for the request phase.

    b. In the acknowledgement phase, the up to k successful inputs contending for an output are acknowledged in a similar manner as the three phase algorithm.

    c. In the send phase, the successful inputs send their packets. At the output of the Batcher network, we use k shuffle exchange networks in parallel for delivering up to k packets per output per slot. The outputs of the Batcher network are connected in a modulo fashion to the inputs of the shuffle exchange networks in the following manner: output 1 to input 1 of network 1, output 2 to input 1 of network 2, . . . , output k to input 1 of network k; output

6.7 Exercises

$k+1$ to input 2 of network 1, ..., output $2k$ to input 2 of network k; output $2k+1$ to input 3 of network 1, ... etc. Prove that the shuffle exchange networks are non-blocking for the k packets.

    d. Compare the throughput and complexity of this scheme with the scheme shown in figure 16 with k staggered parallel switch planes.

10. Very often, the outputs of switching systems are trunk grouped. In other words, a packet can be delivered to any output in a destination trunk group. How can the three phase algorithm be modified for trunk grouping? Specifically, modify the purge mechanism and also provide a mean to assign a trunk address for each packet for which only a trunk group address is given. Assume 2^k outputs form a trunk group. The trunk group address is given by deleting the k least significant bits from the address of any output in the group.

6.8 References

The book by Knuth is a classic on sorting algorithms and networks. The sorting networks proposed by Batcher are often the best discovered so far for practical applications. The bitonic sorters within the Batcher Sorting networks are closely related to banyan networks. This fact was used in our proof. Asymptotically optimal but impractical sorting networks were proposed by Ajtai et. al. Shear sorting as well as more optimal sorting networks are described in the paper by Schnorr et. al.

The use of sorting network for switching was first suggested by Batcher. Huang et. al. were the first to state explicitly the use of cascaded sort-banyan networks for switching. The Starlite switch proposed uses a recirculation mechanism for resolving contention. The three phase mechanism and its variations for resolving contention and for implementing input queued packet switches were proposed by Hui et. al.

The Starlite switch proposed also contains a self-routing copy network. Other self-routing multicast networks were proposed by Turner and Lee. The Appendix describes the construction of Lee.

1. M. Ajtai, J. Komlos and E. Szemeredi, "An $O(n\log n)$ sorting network," *Proc. 15 ACM Symposium on the Theory of Computation*, pp. 1-9, 1983.
2. K. E. Batcher, "Sorting networks and their applications," in *Proc. 1968 Spring Joint Computer Conf.*
3. C. M. Day, J. N. Giacopelli, and J. Hickey, "Applications of self-routing switching to LATA fiber optic networks," in *Conf. Proc. Int. Symp. Switching*, Phoenix, AZ, March 1987.
4. A. Huang, S. Knauer, "Starlite: A wideband digital switch," *Proceeding, Globecom Conference*, Dec. 1988.
5. J. Y. Hui, E. Arthurs, "A broadband packet switch for integrated transport," *IEEE J. on Selected Areas of Comm.*, vol-5, Oct. 1987.
6. D. E. Knuth, *The Art of Computer Programming, Vol. III: Sorting and Searching*, Reading, MA: Addison-Wesley, 1981.
7. T. T. Lee, "Non-blocking copy networks for multicast packet switching," *IEEE J. Selected Areas Commun.*, vol-6, no. 9, Dec. 1988.
8. C. P. Schnorr, A. Shamir, "An optimal sorting algorithm for mesh connected computers," *Proc. 18 ACM Symposium on the Theory of Computation*, pp. 255-262, 1986.
9. J. S. Turner, "Design of a broadcast packet switching network," *IEEE Trans. on Commun.*, pp. 734-743, June 1988.

PART II
Traffic Theory

CHAPTER 7

Terminal and Aggregate Traffic

- 7.1 Finite State Models of Terminals
- 7.2 Modeling of State Transitions
- 7.3 Steady State Probabilities
- 7.4 Superposition of Traffic
- 7.5 Traffic Distribution for Alternating State Processes
- 7.6 Traffic Distribution for Poisson Processes
- 7.7 Broadband Limits and the Law of Large Numbers
- 7.8 Estimating the Traffic Tail Distribution
- 7.9 Appendix- The Improved Large Deviation Approximation

We first focus on modeling terminal traffic. We then examine the superposition of traffic from these terminals. The rest of the chapter evaluates the distribution of superposed random variables and processes. In a broadband environment, we expect a significant degree of traffic aggregation. This inherent advantage for broadband networks is explained using the law of large numbers, which implies that the statistical fluctuations of the aggregate traffic decrease as the number of superposed processes increases. Methods of estimating the tail distribution of the aggregate traffic are also given.

7.1 Finite State Models for Terminals

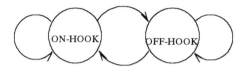

ALTERNATING STATE (ON-OFF) PROCESS

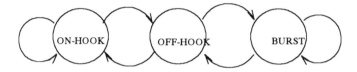

SPEECH WITH SILENCE REMOVAL

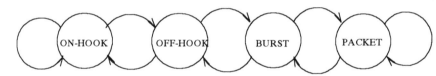

PACKETIZED VOICE

Figure 1. Finite State Models for Telephony

The activity of a terminal very often can be modeled by a finite number of states. We discussed in section 2.4 how the activity of a telephone can be captured by a finite state model. A traditional telephone set can either be in an on-hook (idle) state when it is not in use, or in an off-hook (active) state when someone is making a call. We call this the alternating state process because active periods are always separated by an idle period. If the telephone transmits only when there is speech activity, an additional burst state is needed to describe the terminal. If each burst is further packetized for transmission, the terminal is said to be in a packet transmission state during a transmission. These finite state models of speech for

7.1 Finite State Models for Terminals

telephony are shown in figure 1.

In general, many terminals also exhibit multi-layer activity status, as shown in figure 2. For example, a communication terminal may make a subscription, which consists of a number of calls. Each call may consist of a number of bursts of information. Each burst may be segmented into a number of packets. The layers $l=1,2,3,4$ represent the packet, burst, call, subscription layers respectively, for which the time-scales for these events differ by orders of magnitude. We say layer l is active if the terminal is at a state l or lower. Let us define the load for a terminal at layer l by

$$\rho_l = \text{PROB}(\text{Layer } l \text{ active} | \text{Layer } l+1 \text{ active}) \tag{7.1.1}$$

Let c_1 be the transmission speed of the terminal in the packet state. Strictly speaking, bits are transmitted only when the terminal is in the packet state. However, we can define the average bit rate at layer l by

$$c_l = c_1 \prod_{i=1}^{l-1} \rho_i \tag{7.1.2}$$

Let us illustrate this multi-layer terminal process for telephony with some statistics. A subscription period for telephone service usually lasts for months or even years. Once subscribed, a typical telephone call lasts for a duration of 3 minutes. During busy hours of the day, a telephone is off-hook with a typical probability of $\rho_3 \approx 0.1$. Suppose silent periods are not transmitted by a digital telephone set with a speech activity detector. A typical speech burst lasts for a period of seconds, depending on the algorithm used for detecting speech activity. The burst activity factor is typically $\rho_2 \approx 0.4$. Furthermore, assume that active speech is digitized into a bit stream at 64 kb/s, and segmented into packets of 512 bits for transport over an ATM channel with a capacity of 150 Mb/s. Consequently, each packet is transmitted over a period of 3.4 microseconds, and the load factor of a packetized telephone on an ATM channel during a burst period is $\rho_1 = 4.3 \times 10^{-4}$. Consecutive packets from the same telephone are separated by $1/\rho_1 = 2344$ packet time. In other words, a maximum of 2344 speech bursts can be accommodated at a time on an ATM channel.

Let us now derive the average bit rates at various layers according to equation (7.1.2). At the packet level, we have $c_1 = 150$ Mbit/sec.

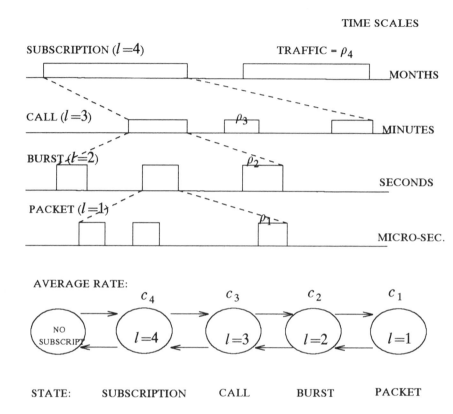

Figure 2. Multi-Layer States of Terminals

The average bit rates at the burst, call, and subscription layers are $c_2 = 64$ kb/s, $c_3 = 25.6$ kb/s, and $c_4 = 2.56$ kb/s.

So far, the state diagram has a linear structure, for which each state is connected to at most 2 states. More complicated models are often employed. For example, a better model for telephony with silence removal can be made if we distinguish various kinds of silence. A speaker may be silent between words in a sentence, pausing for a thought, or listening to the other speaker. Hence various states of silence can be added to the state diagram as shown in figure 3.

7.1 Finite State Models for Terminals

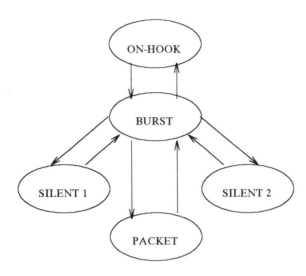

Figure 3. Multiple Silent States for Telephony

For video terminals with motion detection incorporated into the picture encoding algorithm, the terminal may request a high transmission rate during periods of rapid motion, and a lower transmission rate when the amount of motion falls below a certain threshold. If such transmission periods are packetized, we have two packet states associated with the two states of motion respectively. The state diagram is illustrated in figure 4.

Similar to distinguishing various kinds of silence for telephony, it may be advantageous to distinguish various kinds of motion for video communications, such as the movement of people with fixed camera position, the movement of background with moving camera position, or a complete change of scene. The signal encoding algorithm may differentiate these kinds of motion with different bit rates generated for transmission. Very often, an encoding algorithm may simply generate bits at a constantly changing rate, according to the degree of motion and the granularity of the picture. More complicated finite state models have been used to model such fluctuations.

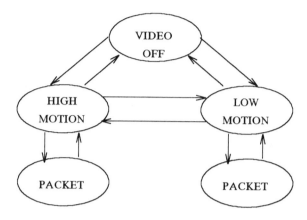

Figure 4. Two Motion States for Video Communications

Terminals can be modeled more accurately if more states are used, at the expense of increased computation complexity for predicting their activities. Given these states, we also have to model how transitions are made from one state to another, which will be described by a probabilistic framework in the next section.

7.2 Modeling of State Transitions

Let us make an abstract model for finite state terminals. Let the states s of a terminal be elements of a finite set S. Over the continuous time horizon $t \geq 0$, the terminal makes transitions from one state, say s, to another state, say s'. Let the random state of a terminal over time be denoted by the random process $S(t)$. $S(t)$ is known as a stochastic process for which $(S(t_1), S(t_2), \cdots, S(t_n))$ has a known probability distribution for all $t_1, t_2, \cdots, t_n$ in time.

A stochastic process is stationary if $S(t)$ is statistically the same as $S(t+t')$ for all t'. In other words, $(S(t_1), \cdots, S(t_n))$ has the same probability distribution as $(S(t_1+t'), \cdots, S(t_n+t'))$ for all $t_1, \cdots, t_n$, and t'.

7.2 Modeling of State Transitions

We define the semi-Markov process as follows. Given a process is in the state s, it would visit the next state s' with probability $f_{ss'}$. Since $f_{ss'}$ is independent of time, the semi-Markov process defined is time-homogeneous. As probabilities, $\sum_{s'} f_{ss'} = 1$. The probability that the process remains in the state s for at most a non-negative period τ before making a transition to s' is denoted by $G_{ss'}(\tau)$. (As a probability distribution, $G_{ss'}(\tau)$ is non-decreasing in τ and equals 1 as $\tau \to \infty$.) Let $g_{ss'}(\tau)$ be the associated probability density function, namely the derivative of $G_{ss'}(\tau)$. Let the mean sojourn time in the state s for the transition $s \to s'$ be $d_{ss'} = \int_0^\infty \tau g_{ss'}(\tau) d\tau$.

An important subclass of the semi-Markov process is the Markov process, which is often assumed for simplifying analysis. Suppose at time t, the process is in state s. A continuous time process is Markov if for all $s, s' \in S$, the state transition $s \to s'$ in the infinitesimal interval $[t, t+\delta t]$ occurs with the infinitesimal probability $p_{ss'} \delta t + o(\delta t)$, for some $p_{ss'} \geq 0$. (As usual, the notation $o(\delta t)$ denotes any function satisfying $o(\delta t)/\delta t \to 0$ as $\delta t \to 0$.) We call $p_{ss'}$ the probability flux from state s to s', which assumes the value 0 if s cannot make a transition to s'. Since $p_{ss'}$ is independent of t, the Markov process defined is considered time-homogeneous. Also $p_{ss'}$ is independent of the time τ the process has spent in the state s. Therefore, Markov processes are considered memoryless because they do not remember the amount of time spent in a state.

Suppose a state s of a Markov process can make a transition only to state s'. What is the probability that the process will stay in state s for more than a period $\tau > T$ from time t? The probability of no transition within each sub-interval of time δt within the interval $[t, t+T]$ is given by $1 - p_{ss'} \delta t - o(\delta t)$. There are $T/\delta t$ such sub-intervals in the interval. The probability of no transition within the interval is given by the product of the probabilities of no transition in each sub-interval. In other words, the desired probability is

$$P(\tau > T) = \left(1 - p_{ss'} \delta t - o(\delta t)\right)^{\frac{T}{\delta t}} \to e^{-p_{ss'} T} \text{ as } \delta t \to 0 \qquad (7.2.1)$$

Therefore the transition considered has an exponential probability distribution for the sojourn time in a state. The probabity density $p(\tau)$ for the sojourn time τ is obtained by differentiating the distribution

$P(\tau \leq T)$ with respect to T, which from eqution (7.2.1) gives $p_{ss'}e^{-p_{ss'}T}$.

A discrete time Markov process is one for which state transitions are allowed only at integer values of t, and that $s(t+1)$, the state at time $t+1$ depends only on the current state $s(t)$ and not the past states $s(t-1)$, $s(t-2)$, $\cdots$, etc. In other words,

$$P(S(t+1)=s_{t+1}|S(t)=s_t, S(t-1)=s_{t-1}, \cdots) \qquad (7.2.2)$$

$$= P(S(t+1)=s_{t+1}|S(t)=s_t)$$

We say that the discrete time Markov process is time-homogeneous if the state transition probability given by the conditional probability $P(S(t+1)|S(t))$ is independent of t. To eliminate the reference to t for time-homogeneous discrete time Markov processes, we denote the state transition probability from state s at time t to state s' at time $t+1$ by $p_{ss'}$. Moreover, $\sum_{s' \in S} p_{ss'} = 1$.

7.3 Steady State Probabilities

We shall be interested mostly in $S(t)$ for large t, when the randomness of state transitions may render $S(t)$ largely independent of the initial condition $s(t=0)$. We say that a process is in steady state when the probability for each state s has converged over time to a probability mass function $P_S(s)=p_s$, independent of the initial state $s(0)$.

It can be shown (see references) that a broad class of finite state processes converges to steady state. For both the continuous time and discrete time processes, each state must be reachable from any other state. Furthermore, the states must be visited non-periodically.

For our purposes, the limiting probabilities can be computed for the following processes.

1. The Markov process

Let us consider a Markov process in steady state. In steady state, making a transition out of a state is as likely as making a transition into that state. In other words, we have the steady state equilibrium equation:

7.3 Steady State Probabilities

$$p_s \sum_{s' \in S} p_{ss'} = \sum_{s' \in S} p_{s's} p_{s'} \qquad (7.3.1)$$

for all $s \in S$. The left hand side is the sum of the probability flow from the state s by the transitions $s \to s'$ and the right hand side is the sum of the probability flow into s by the transitions $s' \to s$. In steady state, these two terms must be equal so that the net probability flow into a state is zero.

For discrete time Markov processes, the same equation (7.3.1) applies. Furthermore, the term $\sum_{s' \in S} p_{ss'}$ in equation (7.3.1) equals 1 because in unit time, the process always makes a transition, including $s \to s$.

2. The semi-Markov process

Consider in steady state the sequence of states visited by a semi-Markov process. This sequence can be viewed as a discrete time Markov process, if we ignore the time durations between transitions. We call the sequence an embedded discrete time Markov process for the corresponding continuous time finite state process. The frequency of occurrence, say q_s, of a state s in the sequence for the embedded Markov process can be found from the discrete time transition probabilities $f_{ss'}$ as shown in the previous paragraph, namely by solving the equilibrium equation

$$q_s = \sum_{s' \in S} f_{ss'} q_{s'} \qquad (7.3.2)$$

for all s, together with the normalization condition $\sum_s q_s = 1$. Given the frequency q_s in the sequence for the embedded Markov process, we can now compute the steady state probabilities p_s for the semi-Markov process. The frequency of occurrence of the consecutive pair ss' in the embedded Markov sequence is given by $q_s f_{ss'}$. The mean duration for the transition is $d_{ss'}$ for this pair. Hence by weighing the frequency of all ss' pairs for given s with the duration $d_{ss'}$, the steady state probability of s for the continuous time semi-Markov process is given by

$$p_s = \frac{\sum_{s'} q_s f_{ss'} d_{ss'}}{\sum_s \sum_{s'} q_s f_{ss'} d_{ss'}} \qquad (7.3.3)$$

3. The alternating state renewal process

An alternating state renewal process is a special case of the semi-Markov process. It has two states named 1 and 2 with no self transitions. Consequently the state transition for the embedded Markov process is $\cdots,1,2,1,2,1,\cdots$ as the name alternating state suggests. Hence $f_1 = f_2 = 0.5$.

Let d_1 and d_2 be the mean sojourn time in the states 1 and 2. By weighing the relative frequencies f_i, $i=1,2$ with the mean durations d_i, the steady state probabilities can be obtained from equation (7.3.3), giving

$$p_1 = \frac{d_1}{d_1+d_2} \tag{7.3.4}$$

$$p_2 = \frac{d_2}{d_1+d_2}$$

The argument generalizes easily for terminals with $n>2$ states for which the states are visited in cycles, namely in the order: $\cdots,n-1,n,1,2,\cdots,n,1,2,\cdots$.

7.4 Superposition of Traffic

We define the traffic process of a terminal as the transmission bit rate requested by a terminal at the interface between the terminal and the network. Suppose a terminal k requires a transmission rate of a_s when it is in the state $s \epsilon S_k$, the set of states for terminal k. We define the random traffic process $R_k(t)=a_s$ if $S_k(t)=s$.

For terminals with packetized transmission, the only state with non-zero a_s is the packet state. For multi-layer terminals, we may ignore the states associated with the smaller time-scales. The traffic process at a particular layer is the average traffic rate at the layer of concern.

Suppose we have K terminals at the same network interface. The aggregate traffic of these terminals is the sum of the individual traffic processes. Very often, we make the assumption that the individual processes are statistically independent. We say two random variables X and Y are independent if the probability $P_{XY}(X=x,Y=y)$ for the

7.4 Superposition of Traffic

joint event $X=x$ and $Y=y$ is equal to $P_X(X=x)P_Y(Y=y)$ for all $x \in X$ and $y \in Y$.

Suppose we look at K terminals in isolation. It is often reasonable to assume that the traffic processes $R_k(t)$ are statistically independent at the same time t or at different times. This independence may not be valid when these terminals interact with each other in conversations.

For a set of K terminals, the superposed traffic is defined as $W(t)=\sum_{k=1}^{K} R_k(t)$. We assume that each terminal k can reach steady state, and that the processes $R_k(t)$ are independent. Hence the process $W(t)$ can also reach steady state. Consequently, the aggregate traffic statistically converges to $W=\sum_{k} R_k$, in which the R_k is the steady state random variable for the random process $R_k(t)$. For simplicity of notation, we denote the distribution of R_k by $P_k(a_k)$ for each rate a_k associated with s_k.

The superposed traffic is a sum of independent random variables. This independency implies that the probability for a particular value of $W=w$ can be expressed as the product of the probabilities of a_k comprising w. In other words,

$$P_W(w) = \sum_{a_1+\ldots+a_K=w} \prod_{k=1}^{K} P_k(a_k) \qquad (7.4.1)$$

In the remainder of this section, we shall look at the superposition of alternating state renewal processes. During the treatment, we shall also introduce some well-known distributions and processes, namely, the Bernoulli distribution, the binomial distribution, the Poisson distribution, as well as the Poisson process.

1. Superposing identical alternating state renewal processes

As shown in the previous section, an alternating state renewal process in steady state can be characterized with steady state probability $p_1=d_1/(d_1+d_2)$ with traffic amplitude 0 in state 1, and probability $p_2=d_2/(d_1+d_2)$ with traffic amplitude say a in state 2. A binary random variable is called a Bernoulli random variable.

In steady state, let us superpose the independently and identically distributed Bernoulli traffic from K terminals. For simplicity, let

$p = p_2$. From equation (7.4.1), the probabilities for W are given by

$$P_W(ja) = \binom{K}{j} p^j (1-p)^{K-j} \tag{7.4.2}$$

because there are $\binom{K}{j}$ ways of having $a_1 + \cdots + a_K = ja$. This distribution is known as the binomial distribution.

2. *Convergence of a binomial distribution to a Poisson distribution*

Suppose we let $K \to \infty$ while keeping $\rho = Kp$ constant. Then the probabilities for W are given by:

$$P_W(ja) = \frac{1}{j!}(1-\frac{1}{K})(1-\frac{2}{K}) \cdots (1-\frac{j-1}{K})(Kp)^j (1-p)^{K-j} \tag{7.4.3}$$

For values of $j \ll K$, each factor in $(1-1/K) \cdots (1-(j-1)/K)$ is effectively 1, while $(1-p)^{K-j} \approx (1-p)^K = (1-\rho/K)^K \approx e^{-\rho}$. Hence

$$P_W(ja) = \frac{\rho^j}{j!} e^{-\rho} + O(\frac{1}{K}) \tag{7.4.4}$$

where the $O(1/K)$ is a polynomial function of $1/K$ with no constant term and $O(1/K) \to 0$ as $K \to \infty$. The distribution in terms of ρ in equation (7.4.4) is known as the Poisson distribution. Thus the superposition of many "bursty" Bernoulli random variables with probability concentrated at 0 produces a Poisson random variable. More generally, the superposition of the traffic processes of many bursty alternating state renewal processes generates a Poisson process to be defined next. Before showing this convergence, let us first study Poisson processes.

3. *Poisson processes*

A Poisson process consists of a number of points (or arrivals) in time. There are two definitions for a Poisson process, one describing the interarrival time (the duration τ between consecutive arrivals) and another describing the arrival probability in a time interval. The first definition states that the interarrival times are independent, and each duration has an exponential distribution, namely $P(\tau > T) = e^{-\lambda T}$. The second definition states that an arrival in an infinitesimal interval $[t, t+\delta t]$ occurs with an infinitesimal probability $\lambda \delta t + o(\delta t)$ independently of other time intervals. These two definitions are congruent since independent infinitesimal arrival probabilities give rise

7.4 Superposition of Traffic

to independent exponential interarrival times as shown in the derivation for equation (7.2.1).

The Poisson arrival process is also related to the Poisson distribution of equation (7.4.4). Let us consider now the number of arrivals in a period T, rather than an infinitesimal period as in the definition for a Poisson process. The number of arrivals m in a period T has a Poisson probability distribution $\frac{(\lambda T)^m}{m!}e^{-\lambda T}$. To prove this, we divide the period T into infinitesimal intervals of duration dt. Suppose the i-th arrival, $1 \leq i \leq m$, arrives in the infinitesimal interval associated with the time t_i, $1 \leq i \leq m$. We have $0 < t_1 < t_2 < \cdots < t_m < T$. The probability of these arrivals in the associated infinitesimal intervals is given by $\prod_{i=1}^{m}(\lambda dt_i)$. Besides these m infinitesimal intervals, all other infinitesimal intervals have no arrival. Using equation (7.2.1), the probability of no arrival in these intervals is $e^{-\lambda T}$. We take the product of these two probabilities, and integrate over all possible $0 < t_1 < \cdots t_m < T$. Hence

$$P(m \text{ arrivals in time } T) \tag{7.4.5}$$

$$= e^{-\lambda T} \int_{t_1=0}^{t_2} \int_{t_2=0}^{t_3} \cdots \int_{t_{m-1}=0}^{t_m} \int_{t_m=0}^{T} \lambda^m \, dt_m \, dt_{m-1} \cdots dt_2 \, dt_1$$

$$= e^{-\lambda T} \lambda^m \frac{T^m}{m!}$$

This distribution is the Poisson distribution of equation (7.4.4), with $\rho = \lambda T$.

4. Convergence of superposed alternating state renewal processes to the Poisson process

Consider the alternating state Markov process shown in figure 5, with Markov transition rate λ from state 1 to 2, and a very large transition rate from state 2 to 1. We now show that this alternating state Markov process generates a Poisson arrival process if we consider the arrival instants at state 1. Since the Markov process spends almost no time in state 2, the time between two consecutive arrivals at state 1 is dominated by the sojourn time in state 1, which is exponentially distributed as shown in equation (7.2.1). Therefore, these arrivals are Poisson in congruence with the definition for the Poisson point process.

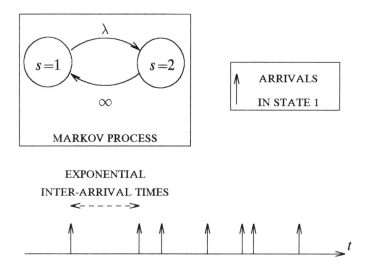

Figure 5. Generating a Poisson Point Process by a Markov Process

Instead of this very bursty (meaning that the fraction of time spent in state 2 is negligibly small) alternating state Markov process, let us consider superposing K alternating state renewal processes operating independently. These processes are not necessarily Markov or identical. Suppose we focus on state 1 for each of these K processes, and assume that the mean time between 2 consecutive arrivals to state 1 is d_k, $1 \leq k \leq K$. In the limit of large K, we show intuitively that the arrival instants for these K processes resemble the Poisson point process for large K.

How long do we have to wait in steady state for the next arrival to state 1 for any of the K processes? This waiting time can be shown to be approximately exponential. Assume that at time $t=0$, the processes are already in steady state. At time t which is substantially less than the individual d_k, the probability that an arrival to state 1 for the

7.4 Superposition of Traffic

process k has occurred is approximately t/d_k. Hence the probability that none of the K processes has an arrival to state 1 within time t is approximated by:

$$\left(1-\frac{t}{d_1}\right)\cdots\left(1-\frac{t}{d_K}\right) \tag{7.4.6}$$

$$= \left(1-\frac{t}{d_1}\right)^{d_1 \frac{1}{d_1}} \cdots \left(1-\frac{t}{d_K}\right)^{d_K \frac{1}{d_K}}$$

$$\approx e^{-t\frac{1}{d_1}} \cdots e^{-t\frac{1}{d_K}}$$

$$= e^{-t\alpha}$$

in which

$$\alpha = \frac{1}{d_1} + \cdots + \frac{1}{d_K} \tag{7.4.7}$$

is the aggregate arrival rate. The probability $e^{-t\alpha}$ is thus the probability that the next arrival time exceeds t. Therefore, the arrival instants for these K processes can be approximated by the Poisson point process.

7.5 Traffic Distribution for Alternating State Processes

In this section, we look at the distribution of the aggregate traffic resulting from superposed alternating state renewal processes in steady state. More specifically, we investigate the distribution of the sum of finitely many random variables. Let us take a second look at the basic equation of the superposition process, namely the product form formula of equation (7.4.1).

Instead of computing $P_W(w)$ directly from (7.4.1), a recursive method known as convolution is often used. Let $w = a_1 + \cdots + a_K$ and $w' = a_2 + \cdots + a_K$. In the process of computing P_W, we may rearrange the indices in equation (7.4.1) to obtain the following convolution formula:

$$P_W(w) = \sum_{a_1} P_1(a_1) \left(\sum_{a_2+\ldots+a_K=w-a_1} \prod_{k=2}^{K} P_k(a_k) \right) \quad (7.5.1)$$

$$= \sum_{a_1} P_1(a_1) P_{W'}(w-a_1)$$

which we define as the convolution $P_1 * P_{W'}$ of the probability distributions. Recursively, the probability distribution of W can be expressed as a convolution of $P_1 * P_2 * \cdots * P_K$.

Define the moment generating function for a random variable W by the following transform on $P_W(w)$

$$\psi_W(s) = \sum_w P_W(w) e^{sw} \quad (7.5.2)$$

Let us transform both sides of equation (7.5.1). Consequently

$$\psi_W(s) = \sum_w \left[\sum_{a_1} P_1(a_1) P_{W'}(w-a_1) \right] e^{sw} \quad (7.5.3)$$

$$= \sum_{a_1} P_1(a_1) e^{sa_1} \sum_{w'} P_{W'}(w-a_1) e^{s(w-a_1)}$$

$$= \psi_{A_1}(s) \psi_{W'}(s)$$

$$= \psi_{A_1}(s) \psi_{A_2}(s) \cdots \psi_{A_K}(s)$$

Therefore, we have the well-known result that the transform of a sum of random variables is given by the product of the transforms of the random variables.

Due to the simplicity of calculating products, it is often computationally more efficient to work in the transform domain than in the convolution domain. Furthermore, the transforms enable us to compute the moments of the random variables efficiently using the following relationship.

$$\frac{d^n \psi_W(s)}{ds^n} \bigg|_{s=0} = \frac{d^n}{ds^n} \sum_w P_W(w) e^{sw} \bigg|_{s=0} \quad (7.5.4)$$

$$= \sum_w P_W(w) w^n e^{sw} \Big|_{s=0}$$

$$= \sum_w P_W(w) w^n$$

Hence we have the name moment generating function for $\psi_W(s)$.

7.6 Traffic Distribution for Poisson Processes

We have shown in section 7.2 that a Poisson process can be generated by superposing a large number of alternating state renewal processes. Instead of generalizing the convolution formula and transform formula of the previous section for this limiting case, we shall proceed directly from the definition of a Poisson process. So let us start by defining a filtered Poisson process.

Let us assume that when the jth Poisson point arrives at time t_j, the traffic generated by the arrival is given by $h(t-t_j)$. Thus, $h(t)$ is the traffic generated by an arrival. We assume that $h(t)$ has finite support, namely that $h(t)$ is nonzero only over a finite amount of time.

The traffic generated by the Poisson point process is then

$$W(t) = \sum_j h(t-t_j) \tag{7.6.1}$$

which is known as a filtered Poisson process since it is obtained by filtering the Poisson point process by the impulse function $h(t)$.

The function $h(t)$ need not be identical for each arrival. Consider superposing many filtered Poisson processes $W_i(t)$ with arrival rate γ_i and impulse response $h_i(t)$. It can be shown that the superposed process $W(t) = \sum_i W_i(t)$ is again Poisson in the sense that the interarrival times are independent and exponentially distributed. The arrival rate is given by $\gamma = \sum_i \gamma_i$. In fact, we may view this superposed process as a Poisson process with rate γ for which an individual pulse happens to be $h_i(t)$ with independent probability γ_i/γ.

Let the time measure of the amplitude a in the function $h_i(t)$ be $b_i(a)$. In other words, the function $h_i(t)$ has the value a for a total duration of $b_i(a)$. To avoid limiting arguments, we assume that $h_i(t)$ are step functions, hence avoiding infinitesimal measures. An illustration of the measure function is shown in figure 6.

Consider $W(t)$ in steady state, with amplitude distribution given by $P_W(w)$. We shall apply an intuitive argument to show that P_W depends only on the aggregate measure function

$$f(a) = \sum_i \gamma_i b_i(a) \tag{7.6.2}$$

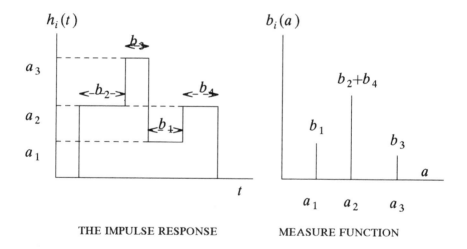

THE IMPULSE RESPONSE   MEASURE FUNCTION

Figure 6. Measure Functions

Proof:

Consider a single pulse $h_i(t)$ with arrival time t_j uniformly distributed in a finite time span. We take a random sample of the amplitude in this time span. Since the sampling time is randomly chosen in the time span, the sampled value does not depend on the shape of the pulse, but rather the measure of a in the pulse. In other

7.6 Traffic Distribution for Poisson Processes

words, we may permute the values of $h_i(t)$ over time t (figure 7) without affecting the sampled value, simply because we chose the sampling instant randomly. Next consider many pulses of shape $h_i(t)$, each with its random and independent arrival time t_j. Since the arrival times are independent for a Poisson process, the multiplicity of pulses sampled at value a has a probability distribution proportional to γ_i. To carry this argument further when we have different kinds of pulses, the multiplicity of pulses sampled at value a depends not so much on the shape of individual $h_i(t)$, but rather the aggregate measure $f(a)$.

□

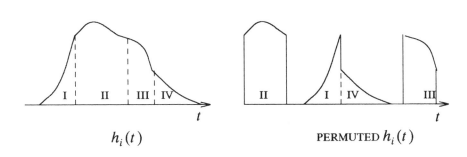

Figure 7. Permuting a Function Without Changing Its Measure

Given this sole dependence on the aggregate measure, we shall without loss of generality assume that $h_i(t)$ are rectangular pulses, with arrival rate γ_i, constant amplitude a_i, and constant duration b_i. Therefore, $f(a)$ assumes the value $\gamma_i b_i$ when $a = a_i$, 0 otherwise. (If a_i and b_i are in fact random, we may split the arrival process with rate γ_i into different Poisson arrival processes with arrival rates weighted by the probability for a_i and b_i.)

Let us consider the probability distribution for W at a random sampling time. Suppose there are n_i pulses of amplitude a_i arriving within the past period b_i. This event occurs with the Poisson probability

$$\frac{(\gamma_i b_i)^{n_i}}{n_i!} e^{-\gamma_i b_i} \qquad (7.6.3)$$

Multiplying this probability for various i and then adding the product for all n_i satisfying $\sum_i n_i a_i = y$, we obtain the probabilities

$$q(y) \stackrel{\Delta}{=} P_W(y) = \sum_{n_i : \sum n_i a_i = y} \prod_i \frac{(\gamma_i b_i)^{n_i}}{n_i!} e^{-\gamma_i b_i} \tag{7.6.4}$$

in which the summation on the right hand side is over all combinations of n_i with the same sum $\sum n_i a_i = y$. This formula is analogous to the product form formula given by equation (7.4.1) for superposed alternating state renewal processes. In section 7.5, we derived a convolution formula for equation (7.4.1). We now derive a convolution formula for superposed filtered Poisson processes, which is surprisingly different from the convolution formula of equation (7.5.3) for sums of finitely many random variables.

Convolution Formula - Filtered Poisson Processes:

$$y\, q(y) = \sum_i \gamma_i a_i b_i\, q(y - a_i) \tag{7.6.5}$$

Using this equation, we can compute $q(y)$ recursively from smaller arguments of y.

Proof

Let us calculate $yq(y)$ using equation (7.6.4) and $y = \sum_j n_j a_j$

$$yq(y) = \sum_{n_i : \sum n_i a_i = y} \left[(\sum_j n_j a_j) \prod_i \frac{(\gamma_i b_i)^{n_i}}{n_i!} e^{-\gamma_i b_i} \right] \tag{7.6.6}$$

Combining the term $n_j a_j$ in the second sum with the term with index $i = j$ in the product, we have a cancellation of terms between n_j in the sum with $1/n_j!$ in the product. By extracting a factor $(\gamma_j b_j)$ from the product, we may express

$$yq(y) \tag{7.6.7}$$

$$= \sum_{n_i : \sum n_i a_i = y} \sum_j \gamma_j a_j b_j \frac{(\gamma_j b_j)^{n_j - 1}}{(n_j - 1)!} e^{-\gamma_j b_j} \prod_{i \neq j} \left[\frac{(\gamma_i b_i)^{n_i}}{n_i!} e^{-\gamma_i b_i} \right] \tag{7.6.7a}$$

7.6 Traffic Distribution for Poisson Processes

$$= \sum_j \gamma_j a_j b_j \sum_{n_i : \sum n_i a_i = y} \frac{(\gamma_j b_j)^{n_j - 1}}{(n_j - 1)!} e^{-\gamma_j b_j} \prod_{i \neq j} \left[\frac{(\gamma_i b_i)^{n_i}}{n_i !} e^{-\gamma_i b_i} \right] \quad (7.6.7b)$$

Using equation (7.6.4), we readily recognize that the inner summation in equation (7.6.7b) is equal to $q(y - a_j)$. Therefore,

$$y q(y) = \sum_j \gamma_j a_j b_j \, q(y - a_j) \quad (7.6.8)$$

which is the desired convolution formula. □

This convolution formula can be used to compute the amplitude distribution of W through recursion. The left hand side of equation (7.6.5) evaluates $q(y)$ in terms, given on the right hand side, of its values at smaller y. Hence a table of the values of $q(y)$ can be evaluated as y increases. We leave the complexity of this evaluation process as an exercise for the reader.

After deriving the convolution formula, we now move on to derive the moment generating function. Using this transform for computing P_W is often more efficient than using the recursive convolution formula. The moment generating function of P_W is given by

$$\psi_W(s) = \sum_y q(y) e^{sy} \quad (7.6.9)$$

There are two methods for deriving $\psi_W(s)$. The first method uses the substitution $q(y)$ given in equation (7.6.4). We leave the derivation as an exercise for the readers. Alternatively, we may derive $\psi_W(s)$ from the convolution formula. Let $\mu_W(s) = \log_e \psi_W(s)$. We prove that for the filtered Poisson process

$$\mu_W(s) = \sum_i \gamma_i b_i \, (e^{sa_i} - 1) \quad (7.6.10)$$

It is noteworthy that the moment generating function (equation (7.5.2)) for the sum of finitely many discrete random variables is a sum of exponentials, whereas the moment generating function for the filtered Poisson process is the exponential of a sum of exponentials.

Proof

Differentiating equation (7.6.9) with respect to s we have

$$\frac{d}{ds}\psi_W(s) = \sum_y y\, q(y) e^{sy} \qquad (7.6.11)$$

Substituting $yq(y)$ in equation (7.6.5) into the right hand side of equation (7.6.11), we have

$$\frac{d}{ds}\psi_W(s) = \sum_i \sum_y \gamma_i a_i b_i q(y - a_i) e^{sy} \qquad (7.6.12)$$

$$= \sum_i \gamma_i a_i b_i e^{sa_i} \psi_W(s)$$

Transposing terms, we have

$$\frac{d\psi_W(s)}{\psi_W(s)} = \sum_i \gamma_i a_i b_i e^{sa_i}\, ds \qquad (7.6.13)$$

which upon integration with the proper integration constant found by setting $\psi_W(0)=1$, yields equation (7.6.10). □

The mean and variance of $W(t)$ are given respectively by differentiating $\psi_W(s)$ in equation (7.6.10). (We use ′ and ″ to represent the first and second derivatives.)

$$E(W) = \psi'_W(s=0) \qquad (7.6.14)$$

$$= \sum_i \gamma_i a_i b_i$$

$$V(W) = \psi''_W(s=0) - (\psi'_W(s=0))^2 \qquad (7.6.15)$$

$$= \sum_i \gamma_i a_i^2 b_i$$

These results for the mean and variance of a filtered Poisson point process are known as Campbell's theorem.

7.7 Broadband Limits and The Law of Large Numbers

In this section, we investigate the properties of the distribution of the sum of a large number n of random variables. We shall show first that the sum, normalized by n should be close to its mean with high probability. This result is known as the law of large numbers.

Applied to broadband networks, the law of large numbers says that the aggregate traffic of a large number of terminals converges to the mean aggregate traffic with high probability. In other words, the statistical fluctuation of the aggregate traffic is averaged out, with diminishing probability for the aggregate traffic to exceed the capacity of a properly engineered transmission link. Consequently, broadband technologies improve network efficiency.

Before proving the law of large numbers, we derive a very useful inequality for bounding the tail distribution of a random variable.

The Markov Inequality

Let Y be a random variable which takes on only nonnegative values. Then for any positive constant d, we have

$$P(Y \geq d) \leq \frac{E(Y)}{d} \qquad (7.7.1)$$

Proof

$$P(Y \geq d) = \sum_{y \geq d} P(y) \leq \sum_{y \geq d} P(y) \frac{y}{d} \qquad (7.7.2)$$

with the last inequality a consequence of $\frac{y}{d} \geq 1$ in the region of summation $y \geq d$. Next, we take the factor $1/d$ outside the summation. We then further overbound the sum by summing over the entire range of Y, thus giving the mean $E(Y)$. □

The Markov Inequality is now used to prove the following.

The Law of Large Numbers

Let X_i, $1 \leq i \leq n$ be independent and identically distributed random variables. Define the random variable

$$S_n = \frac{1}{n} \left(X_1 + \cdots + X_n \right) \qquad (7.7.3)$$

Consider the distribution of S_n for large n. For any positive constants $\epsilon, \delta > 0$, there exists an n_0 such that

$$P_{S_n}(|S_n - E(S_n)| \geq \epsilon) \leq \delta \text{ for all } n \geq n_0 \tag{7.7.4}$$

In other words, we can simultaneously make ϵ and δ small as $n \to \infty$

Proof

Define the random variable $Y = (S_n - E(S_n))^2$. Taking the expectation of Y and using equation (7.7.3), we have

$$E(Y) = \frac{1}{n^2} \sum_i \left[E(X_i^2) - E(X_i)^2 \right] = \frac{1}{n} V(X_i) \tag{7.7.5}$$

in which the variance $V(X_i) = E(X_i^2) - E(X_i)^2$. The last equality is due to the symmetry of X_i for various i. Applying the Markov inequality in equation (7.7.1) to the random variable Y gives

$$P(Y \geq d) \leq \frac{E(Y)}{d} = \frac{V(X_i)}{nd} \tag{7.7.6}$$

Now choose

$$d = \epsilon^2 \quad ; \quad n = n_0 = \frac{V(X_i)}{\epsilon^2 \delta} \tag{7.7.7}$$

Substituting these values into equation (7.7.6), we obtain the result $P(|S_n - E(S_n)| \geq \epsilon) \leq \delta$ for all $n \geq n_0$. □

The law of large numbers is concerned with the sum of n random variables normalized by a factor n, and shows that its probability is concentrated within a small neighborhood of its mean. If we normalize the sum by a smaller factor, we should be able to see more details of the probabilities around the mean. In fact, a normalization factor of $\sqrt{n}$ gives the right scaling factor. If the factor is made slightly larger, the probability plot is squeezed into a narrow region around the mean as $n \to \infty$. If the factor is made slightly smaller, the probability plot is flattened around the mean as $n \to \infty$. When the normalization factor is $\sqrt{n}$, we shall show that the distribution converges to a Gaussian distribution, which is completely specified for given mean and variance. The probability density function of a Gaussian random variable W is given by

7.7 Broadband Limits and The Law of Large Numbers

$$p(W=w) = \frac{1}{\sqrt{2\pi V(W)}} \exp\left(-\frac{(w-E(W))^2}{2V(W)}\right) \qquad (7.7.8)$$

in which $E(W)$ and $V(W)$ are the variance of W.

The Central Limit Theorem

Let X_i, $1 \leq i \leq n$ be independent and identically distributed random variables with finite first and second moments. Define

$$T_n = \frac{1}{\sqrt{n}} \sum_{i=1}^{n} \left(X_i - E(X_i)\right) \qquad (7.7.9)$$

The distribution of T_n converges to the Gaussian distribution with zero mean and variance equal to that for X_i.

The proof of this result is rather involved. The interested reader may consult the textbook by Feller.

7.8 Estimating the Traffic Tail Distribution

The central limit theorem is quite accurate for estimating the distribution of the aggregate traffic within a few standard deviations (the square root of the variance) from the mean. Since the standard deviation grows at a slower rate (proportional to $\sqrt{n}$) than the mean (proportional to n), we can estimate accurately the distribution only within a small neighborhood of the mean as n increases. Consequently for broadband networks, we may apply the Gaussian approximation accurately only for estimating the probability of traffic overflow in the heavy traffic region, where the mean aggregate traffic is only slightly less than the bandwidth provided.

In order to estimate more accurately the probability that the aggregate traffic exceeds a value more than a few standard deviations from the mean, we need the tools of large deviation theory.

We are concerned here with estimating the tail distribution of the aggregate traffic W. This aggregate traffic can result from the superposition of finitely many random variables with moment generating function given by equation (7.5.3), or from the superposition of filtered Poisson processes with moment generating function given by equation (7.6.10). We shall use techniques mostly in the transform domain.

The first approach is closely related to the Markov inequality for the tail distribution.

The Chernoff Bound

$$P(W \geq y) \leq e^{-(s^*y - \mu_W(s^*))} \qquad (7.8.1)$$

in which $\mu_W(s) = \log \psi_W(s)$ is the logarithmic moment generating function. The value s^* satisfies the implicit equation:

$$y = \mu'_W(s) \qquad (7.8.2)$$

Proof

We apply the Markov's inequality of equation (7.7.1) for the random variable e^{sW} and the bound e^{sy}, with $s \geq 0$.

$$P(W \geq y) = P(e^{sW} \geq e^{sy}) \qquad (7.8.3)$$

$$\leq \frac{E[e^{sW}]}{e^{sy}}$$

$$= e^{-(sy - \mu_W(s))}$$

Since s is a free parameter, we may choose the s^* which minimizes the right hand side. Hence

$$P(W \geq y) \leq e^{-(s^*y - \mu_W(s^*))} \qquad (7.8.4)$$

in which s^* is found by setting the first derivative (with respect to s) of the exponent to zero, giving equation (7.8.2). This stationary point is a minimum, which can be shown by proving that the second derivative of $\mu_W(s^*)$ is nonnegative for $s \geq 0$. $\square$

The approximation of equation (7.8.1) can be further sharpened by a factor.

The Improved Large Deviation Approximation

For W being a sum of finitely many random variables or the steady state amplitude of superposed filtered Poisson processes,

$$P(W \geq y) \approx \frac{1}{s^* \sqrt{2\pi \mu''_W(s^*)}} e^{-(s^*y - \mu_W(s^*))} \qquad (7.8.5)$$

7.8 Estimating the Traffic Tail Distribution

Proof

Since the argument is involved, we put it in the Appendix. □

Numerical methods for computing the Chernoff bound and the improved large deviation approximation are also given in the Appendix.

7.9 Appendix- Improved Large Deviation Approximation

We shall prove the large deviation approximation first for filtered Poisson processes.

Recall that N_i is the number of type i bursts, of amplitude a_i arriving during the past period b_i. Assuming Poisson burst arrivals at rate γ_i, we have the following probability density:

$$p_{N_i}(n_i) = \frac{(\gamma_i b_i)^{n_i}}{n_i!} e^{-\gamma_i b_i} \qquad (7.9.1)$$

Since bursts of larger amplitude contribute more significantly to $P(W \geq y)$, it is suggestive to skew the density p_{N_i} according to a_i. Define $N_{s,i}$ as the number of type i burst arrivals, of amplitude a_i, with Poisson arrival rate $\gamma_i e^{sa_i}$ instead of γ_i. Substituting $\gamma_i \to \gamma_i e^{sa_i}$ into equation (7.9.1), we have the skewed density

$$p_{N_{s,i}}(n_{s,i}) = \frac{(\gamma_i e^{sa_i} b_i)^{n_{s,i}}}{n_{s,i}!} e^{-\gamma_i e^{sa_i} b_i} \qquad (7.9.2)$$

$$= p_{N_i}(n_i) \, e^{sn_i a_i - \gamma_i b_i(e^{sa_i}-1)}$$

This equation expresses the relationship between the distributions of $N_{s,i}$ and N_i. Consider the probability densities for $W = \sum_i N_i a_i$ and $W_s = \sum_i N_{s,i} a_i$. Using equation (7.9.2), we obtain the relationship between the distributions of W and W_s.

$$p_W(w) = \prod_{n_i : n_i a_i = w} p_{N_i}(n_i) \qquad (7.9.3)$$

$$= \prod_{n_{s,i}=n_i : \sum_i n_i a_i = w} \left[p_{N_{s,i}}(n_{s,i}) \, e^{-sn_i a_i + \gamma_i b_i (e^{sa_i}-1)} \right]$$

$$= \left[\prod_{n_{s,i}=n_i : \sum_i n_i a_i = w} p_{N_{s,i}}(n_{s,i}) \right] e^{-sw + \sum_i \gamma_i b_i (e^{sa_i}-1)}$$

$$= p_{W_s}(w_s = w) \, e^{-sw + \mu_W(s)}$$

Furthermore, it can be verified easily using equation (7.6.10) that $E(W_s) = \mu'_W(s)$ and $V(W_s) = \mu''_W(s)$.

The desired tail distribution is

$$P(W > y) = \int_y^\infty p_W(w) \, dw \qquad (7.9.4)$$

$$= \int_y^\infty p_{W_s}(w_s) \, e^{-sw_s + \mu_W(s)} \, dw_s$$

$$= e^{-sy + \mu_W(s)} \int_y^\infty p_{W_s}(w_s) \, e^{-s(w_s - y)} \, dw_s$$

We choose $s = s^*$ which satisfies the implicit equation (7.8.2) for the Chernoff bound. Notice that the above integral starts from $y = \mu'_W(s^*)$, the mean of W_{s^*}. To evaluate the integral, we shall use the central limit theorem to approximate $p_{W_{s^*}}$ by the Gaussian distribution, with mean $\mu'_W(s)$ and variance $\mu''_W(s)$. Hence $P(W > y)$ is approximated by:

$$e^{-sy + \mu_W(s^*)} \int_y^\infty \frac{1}{\sqrt{2\pi \mu''_W(s^*)}} \times \qquad (7.9.5)$$

$$\exp\left[-\frac{(w_s - y)^2}{2\mu''_W(s^*)} - s^*(w_s - y) \right] dw_s$$

7.9 Appendix- Improved Large Deviation Approximation

$$= e^{-sy+\mu_W(s^*)} e^{u^2/2} \int_u^\infty \frac{1}{\sqrt{2\pi}} e^{-v^2/2} \, dv \quad ; \quad u = s^*\sqrt{\mu_W''(s^*)}$$

The integral, which is known as the error function, is upper bounded by $\frac{1}{u\sqrt{2\pi}} e^{-u^2/2}$. Hence we have added an extra factor $\frac{1}{u\sqrt{2\pi}}$ for improving the estimate given by the Chernoff bound.

We have proved the large deviation approximation for the filtered Poisson process. The same factor holds for W being a sum of finitely many independent and identically distributed random variables. The proof is similar except at one major point. For the filtered Poisson process, we skew the arrival rate by a factor of e^{sa_i} for Poisson pulse arrivals with amplitude a_i. For sum of random variables X, we skew the probability distribution of X to form X_s such that

$$P_{X_s}(x) = \frac{P_X(x)e^{sx}}{\sum_x P_X(x)e^{sx}} \tag{7.9.6}$$

The rest of the proof is almost identical to that for the filtered Poisson process, and hence is omitted here. □

We now focus on numerical methods for computing the tail distribution of W which is a superposition of filtered Poisson processes and a number of alternating state renewal processes. We've shown that for bursty traffic types modeled by filtered Poisson processes, the logarithmic moment generating function of the sum traffic is given by

$$\mu_W(s) = \sum_i \gamma_i b_i (e^{sa_i} - 1) \tag{7.9.7}$$

whereas for variable rate traffic characterized by the probability versus rate histogram $\{ (a_j, p_j) \}$, the logarithmic moment generating function for the individual call is the usual

$$\mu_k(s) = \log_e \sum_j p_j e^{sa_j} \tag{7.9.8}$$

With mixed Poisson and variable rate traffic, we may sum the logarithmic moment generating functions. The tail distribution of W is tightly approximated by

$$P(W>y) = \frac{1}{s^*\sqrt{2\pi\mu''_W(s^*)}} e^{-(s^*y - \mu_W(s^*))} \tag{7.9.9}$$

where $\mu'_W(s^*) = y$

In order to compute equation (7.9.9), we shall expand the $\mu_k(s)$ in equation (7.9.8) for the individual processes by Taylor series, which can be computed beforehand. The series expansion of equation (7.9.7), which is a sum of exponentials, is straightforward. We may obtain the series expansion of equation (7.9.7) as follows. We first expand the argument to the logarithm into a series. Let

$$\mu_k(s) = \log_e(1 + a_1 s + a_2 s^2 + \ldots) \tag{7.9.10}$$

$$= c_1 s + c_2 s^2 + c_3 s^3 + \cdots$$

It can be shown that we may compute c_i by the following recursive procedure:

$$c_i = a_i - \frac{1}{i} \sum_{j=1}^{i-1} j a_{i-j} c_j \tag{7.9.11}$$

Thus, with these precomputed coefficients, we may easily derive the series expansion of $\mu_W(s)$ as well as its first two derivatives in real time.

Equation (7.9.9) involves solving an implicit equation for s^*, namely setting the first derivative of $\mu_W(s)$ to y. Hence Newton's method can be applied, by using the second derivative of $\mu_W(s)$. Since $\mu_W(s)$ is a positive convex function in s, the root is unique and Newton's method can be very effective. In fact, we may retain the previous values of s^* as well as $\mu_W(s)$ and its derivatives. When $\mu_W(s)$ is changed by superposing another random variable or Poisson process, we may find the perturbation to s^* via Newton's method using the previous derivative. Experience shows that a sufficiently accurate estimate can be obtained with just 1 iterate. The values of $\mu_W(s^*)$ and its derivatives are then updated. The tail distribution of equation (7.9.9) is a simple function of these computed values.

7.10 Exercises

1. Using the moment generating function given in equation (7.6.10), derive Campbell's theorem as stated in equation (7.6.14-15).
2. Since a Poisson process results from the superposition of many bursty processes, we may derive the moment generating function for a Poisson process by taking limits.

    a. Obtain the moment generating function for a Bernoulli random variable with probability ϵ to be a and probability $1-\epsilon$ to be 0.

    b. We superpose K independent Bernoulli random variables, such that $K\epsilon=\rho$. Obtain the moment generating function for the superposed random variable and then obtain the moment generating function for a Poisson process by letting $K\to\infty$. Relate that to equation (7.6.10).

3. Use the substitution in equation (7.6.4) to derive the moment generating function $\psi_W(s)$ for the filtered Poisson process. (Hint: Similar to the derivation of the moment generating function for a sum of finitely many random variable, break up e^{sy} into the components e^{sa_i}. Then interchange the summation and the product.)

4. Estimate the complexity of the convolution methods for computing $P_W(W)$ for all $W\leq y$. Consider W being

    a. The sum of K random variables. Each random variable A_k, $1\leq k\leq K$ may assume only two values $A_k=0$ or $A_k=a_k$.

    b. A filtered Poisson process, with K types of pulses. Each type of pulse has Poisson arrival rate γ_k, amplitude a_k and duration b_k.

5. Consider the Poisson distribution with moment generating function

$$\mu_W(s) = \rho\left(e^{sa}-1\right)$$

Use the improved large deviation approximation given by equation (7.8.5) to obtain the following approximation:

$$P(W \geq y) \approx \frac{1}{\ln\frac{1}{\gamma}} \frac{1}{\sqrt{2\pi\frac{y}{a}}} e^{-\frac{y}{a}[(\gamma-1)-\ln\gamma]}$$

in which $\gamma = a\rho/y$ is a ratio of the mean of the Poisson process (which is equal to $a\rho$ by Campbell's theorem) to the threshold y. Interpret each factor in the above expression. The dominating factor in the above expression is contained in the square brackets in the exponent. Show that this expression is positive for positive γ except when $\gamma=1$. (For $\gamma>1$, the proof given for the Chernoff bound does not hold because $s^* < 0$.)

7.11 References

There is a vast literature on finite state models and analysis of speech traffic, for example, the articles by Brady, Daigle *et. al.*, Heffes *et. al.*, and Minoli. Much less is known about the traffic for signals such as compressed video, for which a discussion can be found in the paper by Maglaris *et. al.* Recently, there is much research performed on traffic issues of packetized voice and video. Many excellent articles can be found in a recent issue edited by Turner *et. al.* The article by Hui also studies traffic integration for data, graphics and full motion video applications through a time-scaled layering of traffic.

Basic probability theory, the Bernoulli and binomial distributions, as well the Poisson, Markov and renewal processes are described in detail in the well-known textbooks by Feller and Cinlar. These distributions and processes are widely used for modeling telephone traffic.

This chapter was structured to illustrate how these distributions and processes are related, particularly in the context of superposed traffic. The convergence of a superposed alternating state renewal process to a Poisson process is well-known in the literature, for example the paper by Sriram *et. al.* While the convolution and transform formulas for the sum of random variables are basic to an undergraduate engineering education, the corresponding formulas for the Poisson process are much less known. The convolution formula was derived (with a simpler proof in our presentation) in the paper by

7.11 References

Gilbert *et. al.* One proof for the transform formula can be found in the textbook by Davenport (chapter 13, exercise 7.2). An efficient method for inverting transforms can be found in the article by Jagerman.

We devoted substantial attention to analyzing rare events and tail distributions, which are also presented in the article by Hui. In the process, analytical tools which will be crucial to the evaluation of blocking probability, the subject for the next chapter, were introduced. Various versions and proofs for the law of large numbers, the central limit theorem, and the large deviation approximations can be found in the textbooks (particularly the second volume) by Feller. Our proof methods followed those found in the textbook by Gallager. The improved estimation of the tail for the filtered Poisson process in the Appendix is an extension of the treatment by Gallager. The numerical methods in the Appendix is new. A proof for the central limit theorem can be found in the textbook by Feller. A simpler but not quite rigorous proof is given in the textbook by Davenport (chapter 12, section 9).

1. P. Brady, "A model for generating on-off speech patterns in two-way conversations," *Bell Syst. Tech. J.*, vol. 48, Sept. 1969.

2. E. Cinlar, *Introduction to Stochastic Processes,* Englewood Cliffs, NJ: Prentice-1975.

3. J. N. Daigle and J. D. Langford, "Models for analysis of packet voice communication systems," *IEEE J. Selected Areas Commun.*, no. 6, Sept. 1986.

4. W. B. Davenport, *Probability and Random Processes,* New York: McGraw Hill, 1970.

5. W. Feller, *An Introduction to Probability Theory and its Application,* Volume I, New York: Wiley. 1968.

6. W. Feller, *An Introduction to Probability Theory and Its Applications,* Volume II, second ed., New York: Wiley. 1971.

7. R. G. Gallager, *Information Theory and Reliable Communication, Chapter 5 and Appendix 5A,* New York: Wiley, 1968.

8. E. N. Gilbert and H. O. Pollak, "Amplitude distribution of shot noise," *Bell Syst. Tech. J.*, March 1960.

9. J. Y. Hui, "Resource allocation for broadband networks," *IEEE J. Selected Areas Commun.*, vol-6 no 9, pp. 1598-1608, Dec. 1988.

10. H. Heffes and D. Lucantoni, "A Markov modulated characterization of packetized voice and data traffic and related statistical multiplexing performance," *IEEE J. Selected Areas Commun.*, no. 6, Sept. 1986.

11. D. L. Jagerman, "An inversion technique for the Laplace transform with application to approximation," *Bell Syst. Tech. J.*, vol. 57, no.3, Mar. 1978.

12. B. Maglaris, D. Anastassiou, P. Sen, G. Karlsson, and J. Robbins, "Performance analysis of statistical multiplexing for packet video sources," *IEEE Trans. on Commun.*, vol. COM-36, July 1988.

13. D. Minoli, "Issues in packet voice communication," *Proceedings of IEE*, vol. 126, Aug. 1979.

14. K. Sriram and W. Whitt, "Characterizing superposition arrival processes in packet multiplexers for voice and data," *IEEE Select. Areas Commun.*, vol SAC-4, Sept. 1986.

15. L. Turner, T. Aoyama, D. Pearson, D. Anastassiou, and T. Minami, Issue on Packet Speech and Video, *IEEE Select. Areas Commun.*, vol SAC-7, June 1989.

CHAPTER 8

Blocking for Single-Stage Resource Sharing

- *8.1 Sharing of Finite Resources*
- *8.2 Truncated Markov Chains and Blocking Probabilities*
- *8.3 Insensitivity of Blocking Probabilities*
- *8.4 The Equivalent Random Method*
- *8.5 Traffic Engineering for Multi-Rate Terminals*
- *8.6 Bandwidth Allocation for Bursty Calls*

The design and management of communication networks involve the allocation of network resources, especially the capacity of transmission links, for transporting the information generated by communication terminals. This chapter focuses on the sharing of transmission links, whereas the same theory can be applied to other problems, such as the sharing of memory and processors.

Blocking of communication on a link occurs when the demand for capacity exceeds the capacity of the link. In general, a connection between two terminals may require the use of many links in tandem. This chapter treats the sharing of a single link and evaluates the probability of blocking, whereas blocking of communication in a multi-stage network will be treated in the next chapter.

In the previous chapter, we looked at models for the terminal traffic process as well as statistics of the aggregate traffic. This aggregate traffic is termed the offered traffic. This chapter furthers the study in the previous chapter by dividing the offered traffic into two parts: the carried traffic and the blocked (or overflow) traffic, as we shall define later.

8.1 Sharing of Finite Resources

In the previous chapter, we defined the aggregate traffic process $W(t) = \sum_k R_k(t)$ as the superposition of independent traffic processes $R_k(t)$ from the terminal k. $W(t)$ converges to a steady state random variable W with a distribution which is a convolution of the distributions of the steady state variables R_k associated with the terminal k.

This aggregate traffic is then offered to say a communication link of finite capacity C for transmission. Hence we also call $W(t)$ the offered traffic. Temporary surges of $W(t)$ beyond the capacity of the link may be dealt with by either delaying the transmission of information, or by dropping information. We call these models the waiting model and the blocking model respectively. In this chapter, we shall deal with the blocking model only, whereas the waiting model will be analyzed in chapter 9 when we study queueing theory.

The blocking model assumes that the overload traffic is lost forever. We shall deal primarily with Poisson arrivals and superposed alternating state renewal processes. For Poisson arrivals (for which an arrival can have different durations and bit rates), the arrival is lost if the additional bit rate makes the offered traffic $W(t)$ exceed the capacity of the link. We call the residue traffic carried by the communication link the carried traffic process $U(t)$. We may also define the overflow traffic process $V(t)$ as $W(t)-U(t)$.

For $R_k(t)$ being an alternating state renewal process, we assume that a state transition from idle (with zero bit rate) to active is immediately renewed in the idle state if the additional offered bit rate makes the offered traffic exceed the capacity of the link. With these conditional arrivals to the active state, let the resulting aggregate

8.1 Sharing of Finite Resources

traffic (which is also the carried traffic) be $U(t)$. The offered traffic process $W(t)$ can look very different from $U(t)$ since the states of the terminals are altered by blocking.

Assuming that the carried traffic can reach steady state, we want to investigate the relation between the steady state random variable U and the traffic processes of the individual terminals. Unfortunately, the terminals are no longer independent since they interact through the sharing of the same resource. However, we can show that the distribution of U is related to W simply by truncating the tail of the distribution of W beyond the bandwidth of the link C for a large class of problems. Let us first define truncated random variables.

Truncated Random Variables

Let W be a discrete real random variable. The random variable U, the truncation of W at C, has distribution given by

$$P_U(u) = \frac{P_W(W=u)}{P_W(W \leq C)} \quad \text{if } u \leq C \tag{8.1.1}$$

$$P_U(u) = 0 \quad \text{if } u > C$$

The denominator normalizes the probability distribution of U to 1. $\square$

Equation (8.1.1) expresses the distribution of U in terms of the distribution of W. Therefore, we can apply the approximation methods for W described in the previous chapter to obtain an approximation for the distribution of U. Specifically, we can apply the large deviation approximation for $P_W(W>C)$ to evaluate the denominator $P_W(W \leq C)$ in equation (8.1.1). The numerator $P_W(W=u)$ can be obtained by the central limit theorem, by the convolution method, or by inverting the moment generating function of W. Very often, the following bounds can be useful as an approximation for the tail distribution (the probability of exceeding a specified value) of U.

Upper and Lower Bounds for Truncated Random Variables

For any $y \leq C$,

$$P_W(y \leq W \leq C) \leq P_U(y \leq U \leq C) \leq P_W(y \leq W) \tag{8.1.2}$$

Proof:

From equation (8.1.1),

$$P_U(y \leq U \leq C) = \frac{P_W(y \leq W \leq C)}{P_W(W \leq C)} \geq P_W(y \leq W \leq C) \qquad (8.1.3)$$

On the other hand,

$$\frac{P_W(y \leq W \leq C)}{P_W(W \leq C)} = \frac{P_W(y \leq W) - P_W(W > C)}{1 - P_W(W > C)} \qquad (8.1.4)$$

$$= P_W(y \leq W) - \frac{P_W(W > C)\left[1 - P_W(y \leq W)\right]}{1 - P_W(W > C)}$$

$$\leq P_W(y \leq W)$$

Hence the inequality (8.1.2). □

Let us apply this truncation to two distributions derived in the previous chapter, namely, the binomial distribution of equation (7.4.2) and the Poisson distribution of equation (7.4.4). The truncated binomial distribution with integral K, u, and C is given by

$$P_U(u) = \frac{\binom{K}{u} p^u (1-p)^{K-u}}{\sum_{i=0}^{C} \binom{K}{i} p^i (1-p)^{K-i}} = \frac{\binom{K}{u} a^u}{\sum_{i=0}^{C} \binom{K}{i} a^i} \qquad (8.1.5)$$

where $a = \dfrac{p}{1-p}$; and $u \leq C$

Using equation (7.4.4), the truncated Poisson distribution is given by

$$P_U(u) = \frac{\dfrac{\rho^u}{u!} e^{-\rho}}{\sum_{i=0}^{C} \dfrac{\rho^i}{i!} e^{-\rho}} = \frac{\dfrac{\rho^u}{u!}}{\sum_{i=0}^{C} \dfrac{\rho^i}{i!}} \quad ; \quad 0 \leq u \leq C \qquad (8.1.6)$$

8.1 Sharing of Finite Resources

Now consider K to be the number of terminals, each modeled by the alternating state renewal process. Assume for each terminal, the mean active period is $1/\mu$ and the mean idle period is $1/\lambda$. Consider the random variable U as the number of active terminals in steady state. These terminals are assumed to share the capacity C (an integer) in a blocking fashion. A major result of this chapter is that the distribution for U is the truncated binomial distribution given by equation (8.1.5), with $a = \lambda/\mu$. If we set $Ka = \rho$ while letting $K \to \infty$, as in the derivation of equation (7.4.4) from equation (7.4.2), then U assumes the truncated Poisson distribution of equation (8.1.6).

In the following section, we shall prove this result for the special case of Markov state transition for the terminals. The general case of the alternating state renewal process is treated in section 8.3.

8.2 Truncated Markov Chains and Blocking Probabilities

In the absence of a resource constraint, the state transition diagram for the number of active terminals has states ranging from 1 to K. With a resource constraint, the state transition diagram for the random variable U is truncated because states with $u > C$ are not allowed. In this section, all state transitions are assumed Markov. The non-truncated and truncated processes are shown in figure 1. We call such a process a birth-death process for the reason that the state can increase or decrease by one at a time.

In chapter 7, we showed that the steady state distribution for the non-truncated Markov birth-death process is binomial. The steady state distribution for the truncated process consequently is a truncated binomial distribution as shown in the following argument. In state $u+1$, the probability density of any one of the $u+1$ active terminals returning to the idle state is given by $(u+1)\mu$. In state u, the probability density of any one of the remaining $K-u$ idle terminals going active is given by $(K-u)\lambda$. In steady state, the net probability exchange between state u and state $u+1$ is zero. Consequently, we have the steady state probability distributions p_u given by

$$p_{u+1} = \frac{(K-u)\lambda}{(u+1)\mu} p_u \quad \text{for } 0 \leq u \leq C-1$$

The same equation is true for all $u < C$ for the non-truncated process. Consequently, it should be obvious why the truncated and non-truncated processes have distributions differing only by a normalization factor. Solving for p_u, we obtain equation (8.1.5), with $a = \lambda/\mu$.

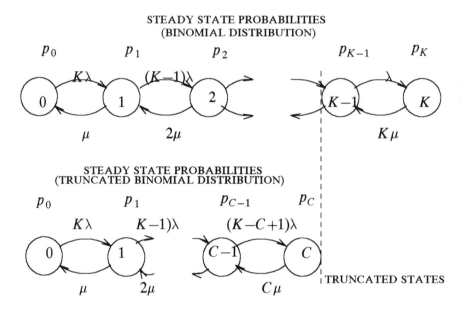

Figure 1. Birth-Death Process for K Terminals

Let us now define the notion of blocking. There are two methods of measuring congestion. The first method measures how fully utilized the resources are, given by the fraction of time that all resources are used. For the birth-death process considered in figure 1, this measure is given by $P_U(U=C)$, namely the probability that the Markov process is at the boundary dictated by the resource constraint $U \leq C$. This measure is also called time congestion or system blocking probability. In the practice of telephony, time congestion can be measured at a telephone trunk group by sampling the group at random times and finding the fraction of sampling times when the trunk group is fully utilized.

The second method measures how often a resource request is denied, given by the fraction of denied resource requests in all

8.2 Truncated Markov Chains and Blocking Probabilities

resource requests. This measure is called call congestion or call blocking for telephony. In the practice of telephony, call congestion can be measured by sampling a telephone at the time of calling and finding the fraction of uncompleted calls. Let us now give an intuitive argument for computing this probability for the birth-death process of figure 1. A more rigorous argument is given in exercise 1. Consider the availability of resources from the point of view of a single terminal. When the terminal becomes active, it is blocked if C out of the remaining $K-1$ terminals are active. The probability of blocking is given by replacing K in equation (8.1.5) by $K-1$, and u by C, giving the well-known Engset formula:

$$B(K;C;a) = \frac{\binom{K-1}{C} a^C}{\sum_{i=0}^{C} \binom{K-1}{i} a^i} \qquad \text{where } a = \frac{\lambda}{\mu} \qquad (8.2.1)$$

In this case, the call congestion and time congestion are approximately the same, except that K is replaced by $K-1$ for call congestion. This difference becomes negligible when K is large.

In this book, we shall deal mostly with time congestion, since it is often easier to compute the steady state probabilities at the boundary of the resource constraint. These two measures are often identical, particularly when the arrival rate of calls at a transmission facility is independent of the state of utilization of the facility. One such example is the Poisson arrival process.

We pointed out in chapter 7 that the Poisson arrival process results from superposing a large number of bursty alternating state renewal processes. Consider the Markov chain in figure 1 and set $K\lambda=\gamma$ while letting $K\rightarrow\infty$. The rate of probability flow from state u to state $u+1$ (up to a fixed state C) becomes a constant γ, while the rate of probability flow from state $u+1$ to state u remains the same $(u+1)\mu$. The resulting Markov birth-death process (figure 2) has the following recursive relationship for the steady state probabilities.

$$p_{u+1} = \frac{\gamma}{(u+1)\mu} p_u \qquad \text{for } 0 \leq u \leq C-1 \qquad (8.2.2)$$

Solving for p_u, we obtain equation (8.1.6) with $\rho=\gamma/\mu$. In particular, the probability that an arriving call is blocked is given by the well-known Erlang blocking formula

$$B(C;\rho) = p_C = \frac{\frac{\rho^C}{C!}}{\sum_{i=0}^{C} \frac{\rho^i}{i!}} \qquad (8.2.3)$$

STEADY STATE PROBABILITIES
(TRUNCATED POISSON DISTRIBUTION)

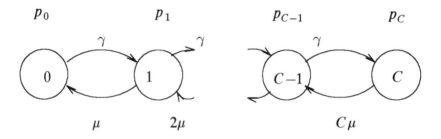

Figure 2. State Space for Poisson Arrivals

We have considered so far that all terminals are of the same type in deriving the blocking probability. We now consider the case of $L>1$ types of terminals, with K_i terminals for type i, $1 \leq i \leq L$. The types may be differentiated by their difference in active time and idle time distributions, or by the bit rate generated while active. Assume Markov state transitions such that each terminal of type i has mean idle time $1/\lambda_i$ and mean active time $1/\mu_i$. We have now a multi-dimensional birth-death process, with each dimension denoting the population n_i, $1 \leq i \leq L$, of active terminals for type i. This Markov chain is illustrated for $L=2$ in figure 3. Without any restriction on the resources consumed, we have a birth-death process in each dimension independent of that for the other dimensions. Hence the joint distribution of the populations of each type is a product form, given by the product of the distributions of the population of each type.

$$P(n_1, \cdots, n_L) = \eta \prod_{i=1}^{L} \binom{K_i}{n_i} \left(\frac{\lambda_i}{\mu_i}\right)^{n_i} \quad ; \text{ for } 0 \leq n_i \leq K_i \qquad (8.2.4)$$

8.2 Truncated Markov Chains and Blocking Probabilities

with η normalizing the total probability to 1.

When we have resource sharing, certain states in the unrestricted Markov chain are not allowed. In this case, we truncate the multi-dimensional Markov chain. Assume that type i terminals generate bits at rate a_i in the active state. The blocked states to be truncated have $\sum_i n_i a_i > C$, in which C is the capacity of the communication link. Consequently, the boundary of the blocked states is linear as shown in figure 3. We shall show that the steady state probability of each state in the truncated multi-dimensional Markov chain is the same as that for the unrestricted Markov chain, except for a normalization factor.

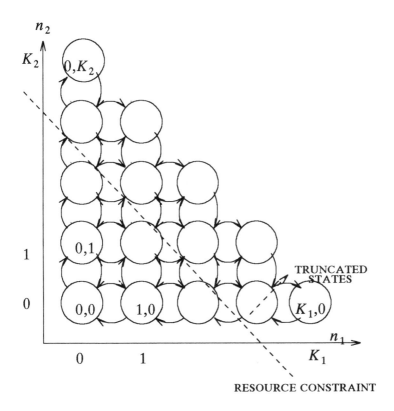

Figure 3. Two Dimensional Markov Birth-Death Processes

Proof:

We examine the equilibrium equation more carefully. The equilibrium equation states that the net probability flowing into a state s is equal to the probability flowing out of that state in statistical equilibrium, namely

$$\sum_{s'} p_{s'} p_{s's} = p_s \sum_{s'} p_{ss'} \qquad (8.2.5)$$

A stronger condition of balance than equation (8.2.5) maintains the balance of flow from state s to s' and the flow from state s' to s:

$$p_{s'} p_{s's} = p_s p_{ss'} \qquad (8.2.6)$$

This equation is known as the detailed balance equation. Satisfaction of equation (8.2.6) for all states s automatically implies the satisfaction of equation (8.2.5), shown by summing equation (8.2.6) over s'.

A birth-death process (figure 1), which is a one-dimensional Markov chain, satisfies detailed balance since probability flow between adjacent states must be balanced in steady state (see equation (8.2.2) for example). Consider the states in the non-truncated multi-dimensional Markov chain with identical n_i for all i except for one particular index j. These states constitute a one-dimensional Markov birth-death process. Detailed balance is maintained in this restricted one-dimensional Markov process, since equation (8.2.4) is a product form for the probabilities of one-dimensional Markov processes.

Suppose we remove the pair of transitions $s \rightarrow s'$ and $s' \rightarrow s$ from the non-truncated multi-dimensional Markov chain. Since the pair is balanced, its removal would not disturb the steady state probabilities of the states connected by the pair. If we remove all transition pairs of a state, the state becomes isolated and has zero steady state probability. However, the steady state probabilities for the other states remain unchanged, except for a renormalization factor taking into account the zero probability of the isolated state. Therefore, a truncated multi-dimensional Markov chain with detailed balance differs from the unrestricted multi-dimensional Markov chain only by a normalization factor. $\square$

Consider taking the Poisson limit by letting $K_i \rightarrow \infty$ while keeping $K_i \lambda_i = \gamma_i$ constant. From equation (8.2.4), we have the following distribution for the carried traffic

8.2 Truncated Markov Chains and Blocking Probabilities

$$P_U(u) = \eta \sum_{n_i : \sum n_i a_i = u} \prod_{i=1}^{L} \frac{\rho_i^{n_i}}{n_i!} \qquad (8.2.7)$$

with $\rho_i = \gamma_i/\mu_i$. As expected, this truncated distribution has a similar form to the amplitude distribution for the Poisson process W, given in equation (7.6.4). (The exponential factor in equation (7.6.4) is absorbed into the constant η). Therefore, the techniques for estimating the distribution for W can also be used to estimate the distribution for U, as well as the normalization factor η.

The time congestion for type i terminals is defined by

$$P_{B,i} = P_U(U > C - a_i) \qquad (8.2.8)$$

Therefore, terminals with a larger a_i suffer a higher probability of blocking. Estimates of the difference in blocking for multi-rate terminals are given in the exercise.

8.3 Insensitivity of Blocking Probabilities

Instead of exponentially distributed time distributions for the Markov process, we now consider the alternating state renewal process with general active and idle time distributions. The steady state distribution of the number of active terminals is again truncated binomial, independent of the detailed shape of the active time distribution or idle time distribution except through the ratio of their means, if the system can reach steady state. Not all such systems can reach steady state though. For example, if all terminals have active and idle times equal to the same constant, the behavior of the system is periodic. Hence the system cannot reach steady state.

A probability density function for time with a finite number of discontinuities may be approximated by a finite set of values $\{t_i\}$. The time t_i occurs with probability p_i as shown in figure 4. As we make the size of the finite set as large as we please, the original density function is closely approximated. We study the properties of the system using this discrete approximation, and then deduce the behavior for the original distribution by taking limits.

The random time distribution is simulated by an object sojourning in one of the shift registers, each associated with a t_i. The object branches randomly into the shift register i with probability p_i as shown in figure 4. Let us further discretize the unit time into M slots. Hence the ith shift register has Mt_i (rounded to the nearest integer) positions. The object is shifted one position per slot time. After the object is shifted through a register, it departs from the stage of parallel shift registers. We now give an elementary and intuitive proof for the truly basic result of insensitivity.

Insensitivity Theorem for Superposed Alternating State Processes

Let K alternating state renewal processes, each of mean active time $1/\mu_k$ and mean idle time $1/\lambda_k$ for $1 \leq k \leq K$, share a transmission capacity of C. In the active state, the process generates bits at a rate of a_k. Let $c_k = 1$ if the process k is active, $c_k = 0$ otherwise. Then the steady state probabilities for the c_k are given by

$$p_{c_1 \cdots c_K} = \eta \prod_{k=1}^{K} \left(\frac{\lambda_k}{\mu_k} \right)^{c_k} \quad \text{if} \quad \sum_{k=0}^{K} c_k a_k \leq C \tag{8.3.1}$$

$$p_{c_1 \cdots c_K} = 0 \quad \text{otherwise}$$

in which η is the normalization constant. Notice that equation (8.3.1) is a special case of equation (8.2.4) with $K_i = 1$, which was derived for the Markov process.

Proof:

We use two stages of shift registers each simulating the active and idle periods as shown in figure 5 for the process k. The terminal is modeled by an object shifting through the first (active) stage, and then branches into the second (idle) stage. Let us assume that there are m_k possible values of active time, each of $M\alpha_{ik}$ slot times, $1 \leq i \leq m_k$. The branching probabilities for these active times are f_{ik}, $1 \leq i \leq m_k$. Likewise, let us assume that there are n_k possible values of idle time, each of $M\beta_{ik}$ slot times, $1 \leq i \leq n_k$. The branching probabilities for these idle times are g_{ik}, $1 \leq i \leq n_k$. When the object is shifted out of a second stage register, the object returns immediately to the first stage. However, the object immediately enters the second stage through the overflow branch if the terminal is blocked. This happens when the condition in equation (8.3.1) is violated.

Consider how these K terminals with their two-stage simulations interact with each other. Let us introduce an inconsequential artifact of having these K 2-stage simulations shifted one after another and

8.3 Insensitivity of Blocking Probabilities

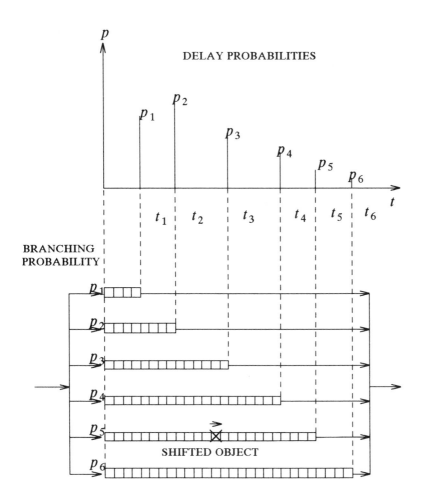

Figure 4. Simulating a Time Distribution by Shift Registers

cyclically. In other words, we shift process k for 1 slot, then process $k+1, \ldots$, process K, and then back to process 1. In the process of shifting the object k out of the second stage, it is allowed to enter the

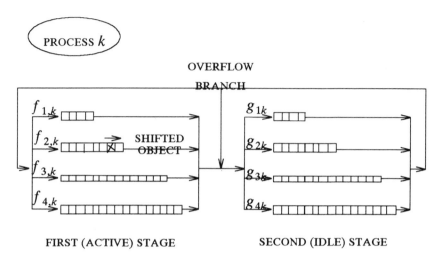

Figure 5. Modeling Blocking by a Semi-Markov Process

first stage only if the resource constraint condition of equation (8.3.1) is not violated after the transition is made. Otherwise, the object re-enters the second stage immediately through the overflow branch.

Let us define discrete time Markov states for these K simulations. Let the state s_k for the process k be the triple (c_k, d_k, e_k) defined as follows. As defined before, $c_k = 1$ if the object is in the first (active) stage, and $c_k = 0$ if the object is in the second (idle) stage. Let the object be in the d_k-th register in the respective stage. Let e_k be the position in the respective register. Furthermore, we define the state s as the vector $(s_1, \cdots, s_K)$. The steady state probability p_s can be found by the balance of flow equations if the discrete time Markov process is aperiodic.

For each state s, let us define the function

$$h_k(s) = \begin{array}{ll} f_{d_k k} & \text{if } c_k = 1 \\ g_{d_k k} & \text{if } c_k = 0 \end{array} \tag{8.3.2}$$

Simply speaking, $h_k(s)$ gives the branching probability of the register containing k. Let us make the hypothesis that the steady state probabilities for s are given by

8.3 Insensitivity of Blocking Probabilities

$$p_s = \eta \prod_{k=1}^{K} h_k(s) \quad \text{if } \sum_{k=1}^{K} c_k a_k \leq C \tag{8.3.3}$$

$$p_s = 0 \quad \text{otherwise}$$

Consider $s' \to s$ in the process of shifting the process k. Equation (8.3.3) implies the following relationship:

$$\frac{p_s}{h_k(s)} = \frac{p_{s'}}{h_k(s')} \tag{8.3.4}$$

We shall show that equation (8.3.3) satisfies the equilibrium equation for the discrete time Markov process, namely

$$p_s = \sum_{s'} p_{s's} p_{s'} \tag{8.3.5}$$

for the transitions $s' \to s$. There are two cases to consider when process k is shifted:

1. $e_k > 1$, which means that the object is shifted within a register. Hence s came from a unique s', with $p_{s's} = 1$. Substituting $p_{s's} = 1$ and $p_{s'}$ in equation (8.3.4) into equation (8.3.5) confirms our hypothesis for p_s for this case.

2. $e_k = 1$, which means that the object k is shifted according to one of the following 3 cases:

   A. From the first stage to the second stage,

   B. From the second stage to the first stage if there is no blocking, or

   C. From the second stage to the second stage if there is blocking.

We can determine uniquely which of the above cases occurred if s, the current state after the transition, is given. There are many possible previous states s', with $p_{s's} = h_k(s)$.

For the above three cases, let us substitute $p_{s'}$ in equation (8.3.4) and $p_{s's} = h_k(s)$ into equation (8.3.5):

$$p_s = \sum_{s'} h_k(s) \frac{p_s}{h_k(s)} h_k(s') = p_s \sum_{s'} h_k(s') = p_s \tag{8.3.6}$$

hence confirming again our hypothesis for p_s.

Let us now look at the unconstrained process with no overflow branch. The steady state probabilities for s are identical to that given in equation (8.3.3) except for a different normalization constant η' instead of η, and the absence of the resource constraint $\sum c_k a_k \leq C$. This can be verified by the same argument used for the constrained process, except that case C considered above is absent.

We now sum p_s over all s with identical c_k, but possibly different d_k and e_k, for $1 \leq k \leq K$. Naturally, the resulting probabilities $p_{c_1 \cdots c_K}$ for the constrained process differ from those for the unconstrained process by the same constant. For the unconstrained process k (which is an alternating state renewal process), the steady state probabilities are given by equation (7.3.4), namely

$$p_{c_k=0} = \frac{\frac{1}{\lambda_k}}{\frac{1}{\lambda_k} + \frac{1}{\mu_k}} \quad ; \quad p_{c_k=1} = \frac{\frac{1}{\mu_k}}{\frac{1}{\lambda_k} + \frac{1}{\mu_k}} \qquad (8.3.7)$$

Hence for the K unconstrained processes, we have

$$p_{c_1 \cdots c_K} = \prod_{k=1}^{K} p_{c_k} = \eta' \prod_{k=1}^{K} \left(\frac{\lambda_k}{\mu_k}\right)^{c_k} \qquad (8.3.8)$$

after absorbing constant terms into η'. Hence the steady state probabilities for the constrained process given in equation (8.3.1) follow. This completes the proof.

Alternatively, we may also compute $p_{c_1 \cdots c_K}$ for the constrained process directly from equation (8.3.3). Let us form the macrostate σ defined as the set of states s with the same c_k and d_k for each k. The number of s in σ is given by the product of the number of positions of the registers containing an object. Each of these s has the same probability, namely $\prod_k h_k(s)$. Hence

$$p_\sigma = \eta \left(\prod_{k:c_k=1} M\alpha_{d_k k} f_{d_k k}\right) \left(\prod_{k:c_k=0} M\beta_{d_k k} g_{d_k k}\right) \qquad (8.3.9)$$

8.3 Insensitivity of Blocking Probabilities

Let us now form larger macrostates v defined as the set of σ with the same c_k for each k. We sum the steady state probabilities of equation (8.3.9) over the m_k registers for the first stage, and n_k registers for the second stage. This sum of products can be rearranged to become a product of sums

$$p_v = \eta \left(\prod_{k:c_k=1} M \sum_{i=1}^{m_k} \alpha_{ik} f_{ik} \right) \left(\prod_{k:c_k=0} M \sum_{i=1}^{n_k} \beta_{ik} g_{ik} \right) \quad (8.3.10)$$

We readily realize that the sums in the above expression are the mean holding time $1/\mu_k$ and the mean idle time $1/\lambda_k$ respectively. Multiplying the right hand side by $\lambda_1 \cdots \lambda_K$, and then absorbing M into the normalization constant η, we can express this steady state probability as

$$p_{c_1 \cdots c_K} = \eta \prod_k \left(\frac{\lambda_k}{\mu_k} \right)^{c_k} \quad (8.3.11)$$

which is equation (8.3.1). □

The insensitivity of the Engset blocking formula to the detailed active and idle time distributions follows immediately, since the binomial distribution is obtained by superposing identical alternating state renewal processes. Similarly, the insensitivity of the Erlang blocking formula to the detailed active time distribution follows from the fact that the Poisson distribution is the limiting case for the binomial distribution.

8.4 The Equivalent Random Method

So far, simple solutions have been found for the steady state probabilities of capacity utilization for terminals modeled by the alternating state renewal process and the Poisson process. Unfortunately, an exact solution cannot be obtained for most other processes. One example is the overflow traffic $V(t) = W(t) - U(t)$ defined in section 8.1 for the Poisson process $W(t)$. Unlike the Poisson process $W(t)$ for which the arrival instants are homogeneous in time, overflow arrivals occur in clusters during periods of congestion.

8. Blocking for Single-Stage Resource Sharing

An attempt to approximate blocking for these traffic processes models the traffic arrivals as Poisson, with matched first and second moments. Consequently, we may still use the Engset and Erlang blocking formula for the probability of blocking. This technique, known as the equivalent random method, has been very useful for a host of traffic engineering problems.

Suppose the mean α and variance v of the offered traffic are known. For the equivalent Poisson arrival process, we want to find the probability of blocking of an "arrival". What do we mean by an arrival? Here we have to introduce the notion of scaling. This scaling measures the peakedness (or the lumpiness) of the traffic.

Consider the following example. Suppose we have Poisson arrivals of calls requiring 1 unit of capacity with constant holding time b. Let the arrival rate be λ, consequently the offered traffic is $\rho = \lambda b$. The mean and variance of the offered traffic, by Campbell's theorem in equations (7.6.14-15), are both equal to ρ. Now suppose we have instead the same arrival rate and holding time for calls requiring c units of bandwidth. By Campbell's theorem, the mean and variance of the offered traffic now become $c\rho$ and $c^2\rho$ respectively. The blocking probability of calls requiring c units of bandwidth sharing the capacity C is the same as that for calls requiring 1 unit of bandwidth sharing a capacity of C/c (assuming c divides C). This blocking probability is given by the Erlang blocking formula $B(C/c, \rho)$.

From this example, we see that the scaling factor c defines the granularity of the "representative" call in the offered traffic. Therefore, the shared bandwidth should be scaled by this granularity. In the above example, dividing the variance by the mean gives $c^2\rho/c\rho = c$, the scaling factor. It should be noted that the holding time has no effect on the granularity, as is evident from the fact that increasing the holding time increases the mean and variance by the same factor by Campbell's theorem.

For the more general case of mean α and variance v, we have the scaling factor, also known as peakedness, given by $c = v/\alpha$. In the blocking formula $B(C/c, \rho)$ above, the offered traffic ρ is given by the mean $c\rho$ divided by the scaling factor c. Hence the effective offered traffic for the equivalent Poisson process with mean α is given by α/c. Therefore, the equivalent blocking probability is given by

$$P_B = B\left(\frac{C}{c}, \frac{\alpha}{c}\right) \quad ; \text{where } c = \frac{v}{\alpha} \quad (8.4.1)$$

8.4 The Equivalent Random Method

This formula is known as the Hayward approximation formula.

In the following, we shall apply the equivalent random method for two applications.

1. Multi-rate traffic with Poisson arrivals

In section 8.3, we derived the product form solution for the steady state probabilities of arrival processes with multiple classes of bit rates, arrival rates, and holding time distributions. Also, large deviation theory can be applied for approximating the blocking probability. Here, we use the equivalent random method as an alternative approximation method.

By Campbell's theorem, the superposed filtered Poisson process has mean and variance given in equations (7.6.14-15). Hence we can apply equation (8.4.1) for calculating the blocking probability of a "representative" call. This blocking probability should be quite uniform among calls of different rates provided that the available bandwidth C is much larger than any call rate. More accurately, we may approximate the blocking probability of calls with bit rate a_i by

$$P_{B,i} = B\left(\frac{C-a_i}{c}, \frac{\alpha}{c}\right) \quad ; \text{ where } c = \frac{v}{\alpha} = \frac{\sum_i \gamma_i a_i^2 b_i}{\sum_i \gamma_i a_i b_i} \quad (8.4.2)$$

in which γ_i, a_i, b_i are respectively the arrival rate, amplitude, and mean holding time for call i. If $(C-a_i)/c$ is not an integer, it has to be rounded off in order to compute $P_{B,i}$.

2. The traffic overflow process

Consider arrivals of calls with unit bandwidth requirement, exponential interarrival time distribution of mean $1/\lambda$ and exponential service time distribution of mean $1/\mu$. This traffic $\rho = \lambda/\mu$ is offered first to a primary transmission facility with a capacity for C calls. We want to find the mean and the variance of the overflow traffic. The overflow traffic is offered to a secondary transmission facility which we assume temporarily to be of infinite capacity.

Let i be the number of calls using the primary facility and j be the number of calls using the secondary facility. If we arrange the states into the dimensions i and j, we have the state transition diagram given in figure 6. In the diagram, a transition to a higher j can only occur when $i=C$, the capacity of the primary facility.

The steady state probabilities of this state transition diagram can be found after a tedious derivation. We shall omit the derivation and cite the first two moments of the overflow traffic instead.

The mean is given by

$$\alpha = \rho\, B(C,\rho) \qquad (8.4.3)$$

This is intuitively obvious since overflow can only arrive during the blocking state which occurs with probability $B(C,\rho)$. The variance is given by

$$v = \alpha\left(1-\alpha+\frac{\rho}{C+1-\alpha-\rho}\right) \qquad (8.4.4)$$

A derivation of this result is complicated, and can be found in the book by Cooper, section 4.3.

Suppose we have a number of primary transmission facilities, for which the overflow calls from each primary facility are offered to a common secondary facility of finite capacity as shown in figure 7. When we allow several overflow traffics to share the secondary facility, the aggregate overflow traffic has mean and variance found by simply adding the means and variances of the overflow streams. The probability of blocking for the secondary facility can be found by using the Hayward approximation formula (8.4.1).

8.5 Traffic Engineering for Multi-Rate Terminals

In the previous sections, we analyzed the probability of blocking for a given offered traffic. In this section, our primary concern is the inverse problem: What is the maximum offered traffic for not exceeding a given probability of blocking? We call this problem the traffic engineering problem.

Consider the multi-rate Poisson traffic process of section 8.2. For convenience, we restate the traffic model here. We have a mixture of L types of terminals sharing a common bandwidth C. A terminal of type i generates calls of bit-rate a_i for a mean duration b_i. Furthermore, we assume the arrival of calls for type i terminals is Poisson of rate γ_i, generating a traffic of $\rho_i = \gamma_i b_i$. The probabilities

8.5 Traffic Engineering for Multi-Rate Terminals

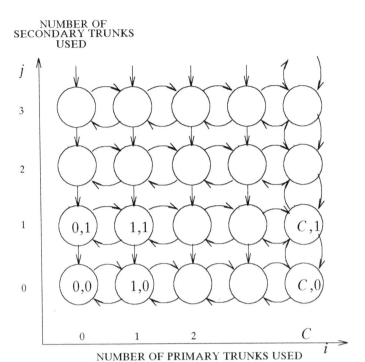

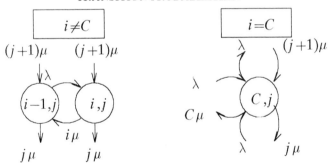

Figure 6. Two Dimensional Markov States for Traffic with Overflow

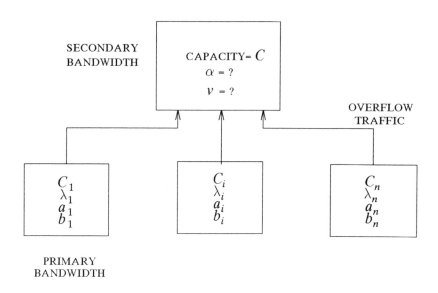

Figure 7. Using Secondary Transmission Facility for Overflow

for the carried traffic U is given by equation (8.2.7). We want to find the set of ρ_i such that the call blocking probabilities $P_{B,i}$ given in equation (8.2.8) is within a prescribed bound for each terminal type i. By limiting the number of terminals of each type, the offered traffic ρ_i is controlled.

For simplicity, we may assume the same bound P_B for all $P_{B,i}$. The set of such ρ_i such that $P_{B,i} \leq P_B$ is called $R(P_B)$, which is shown by the curved region in figure 8. To simplify the analysis further, we shall approximate $P_{B,i}$ by $P_W(W>C)$, instead of using $P_{B,i} = P_U(U>C-a_i)$ given in equation (8.2.8). This approximation does not affect the region $R(P_B)$ significantly. Furthermore, we shall focus on using the Chernoff bound for computing $P_W(W>C)$. We choose the Chernoff bound method for two reasons. First, the resulting estimate $P(W>C)$ is also an upper bound on the probability of blocking, hence the resulting $R(P_B)$ turns out to be conservative. Second, the Chernoff bound gives several appealing economic interpretations as we shall see next.

8.5 Traffic Engineering for Multi-Rate Terminals

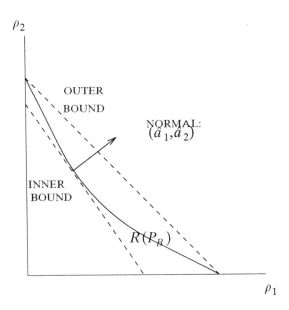

Figure 8. The Region $R(P_B)$

We shall focus on the exponent of the Chernoff bound. In other words, we evaluate the logarithm of $P_W(W>C)$ at the boundary of $R(P_B)$. The Chernoff bound derived in chapter 7 is restated here:

$$\ln\frac{1}{P_B} = \max_s \left[sC - \mu_W(s) \right] \qquad (8.5.1a)$$

$$= s^*C - \mu_W(s^*) \qquad \text{with } \frac{\partial}{\partial s}\mu_W(s^*) = C \qquad (8.5.1b)$$

(For convenience, the bound is treated as an equality). For a given P_B, the above equation gives a relationship for the ρ_i for the boundary of $R(P_B)$.

The parameter s^* in equation (8.5.1b) has the interesting interpretation of being the shadow price (measured by the logarithmic blocking probability) for the bandwidth C shown by differentiating equation (8.5.1b) with respect to C.

$$\frac{\partial}{\partial C}\ln\frac{1}{P_B} = \frac{\partial}{\partial C}\left(s^*C - \mu_W(s^*)\right) \tag{8.5.2}$$

$$= s^* + \frac{\partial s^*}{\partial C}\left[C - \frac{\partial}{\partial s}\mu_W(s^*)\right]$$

$$= s^*$$

in which the second equality results from applying the chain rule for differentiation, and the third equality is a consequence of the implicit equation for s^* in equation (8.5.1b).

The logarithmic moment generating function for Poisson traffic was given by equation (7.6.10):

$$\mu_W(s) = \sum_i \rho_i \left(e^{sa_i} - 1\right) \stackrel{\Delta}{=} \sum_i \rho_i \, \hat{a}_i(s) \tag{8.5.3}$$

It is noteworthy that $\mu_W(s)$ is a linear function of ρ_i and $\hat{a}_i$. We call $\hat{a}_i(s^*)$, a function of s^* and a_i, the relative cost of type i traffic for reasons to be shown later. First, we shall show that this linearity implies that $R(P_B)$ is convex, as shown in figure 8.

Proof:

Consider two points $\underline{\rho}' = (\rho_1', \cdots, \rho_L')$ and $\underline{\rho}'' = (\rho_1'', \cdots, \rho_L'')$. Consider the right hand side of equation (8.5.1) for the linear combination $\underline{\rho} = \alpha\underline{\rho}' + (1-\alpha)\underline{\rho}''$.

$$\max_s \left(sC - \sum_i \rho_i \hat{a}_i(s)\right) \tag{8.5.4}$$

$$= \max_s \left[\alpha(sC - \sum_i \rho_i' \hat{a}_i(s)) + (1-\alpha)(sC - \sum_i \rho_i'' \hat{a}_i(s))\right]$$

$$\leq \alpha \max_s \left[sC - \sum_i \rho_i' \hat{a}_i(s)\right] + (1-\alpha)\max_s \left[sC - \sum_i \rho_i'' \hat{a}_i(s)\right]$$

8.5 Traffic Engineering for Multi-Rate Terminals

$$= \alpha \ln \frac{1}{P_B{'}} + (1-\alpha) \ln \frac{1}{P_B{''}}$$

in which $P_B{'}$, $P_B{''}$ are the resulting values of the optimizations.

Consider $P_B{'} = P_B{''} = P_B$, namely that both $\underline{\rho}'$ and $\underline{\rho}''$ are on the boundary of $R(P_B)$. Blocking at $\underline{\rho}$ is larger than P_B as a result of equation (8.5.4), hence $\underline{\rho}$ is outside $R(P_B)$. Therefore, the boundary for $R(P_B)$ is convex. $\square$

This convexity property provides convenient inner and outer bounds for the region $R(P_B)$. An outer bound is given by first computing the vertices of the region with $\rho_i = 0$ except for one ρ_i. These k vertices define a plane forming an outer bound for $R(P_B)$.

On the other hand, any tangent plane to the boundary of $R(P_B)$ is an inner bound for $R(P_B)$. Let us derive the formula for these tangent planes. Substituting equation (8.5.3) into equation (8.5.1) and then taking partial derivatives with respect to ρ_i,

$$\frac{\partial}{\partial \rho_i} \ln \frac{1}{P_B} = \frac{\partial}{\partial \rho_i} \left[s^* C - \sum_j \rho_j \hat{a}_j(s^*) \right] \qquad (8.5.5)$$

$$= \frac{\partial s^*}{\partial \rho_i} \left[C - \sum_j \rho_j \frac{\partial}{\partial s^*} \hat{a}_j(s^*) \right] - \hat{a}_i(s^*)$$

$$= - \hat{a}_i(s^*)$$

with the second equality a result of the chain rule for differentiation, and the third equality resulting from the implicit equation defining s^* in equation (8.5.1). Consequently, $\hat{a}_i(s^*)$ is the marginal cost of degradation of the grade of service (measured by the logarithmic blocking probability) for adding traffic for terminals of type i. Furthermore, the vector $\underline{\hat{a}}(s^*) = (\hat{a}_1(s^*), \cdots, \hat{a}_L(s^*))$ is a normal to the tangent plane at the boundary of $R(P_B)$ as shown in figure 8, since the differential vector $\underline{d\rho} = (d\rho_1, \cdots, d\rho_L)$ in the tangent plane satisfies

$$\sum_i \left[\frac{\partial}{\partial \rho_i} \ln \frac{1}{P_B} \right] d\rho_i = 0 \qquad (8.5.6)$$

Hence by the substitution from equation (8.5.5), we have

$$\underline{\hat{a}}(s^*)\, d\underline{\rho}^t = 0 \qquad (8.5.7)$$

in which t denotes the transpose of a vector. Hence $d\underline{\rho}$ and $\underline{\hat{a}}(s^*)$ are orthogonal to each other. Interpreted alternatively, $\bar{\hat{a}}_i(s^*)$ represents the price per unit traffic for type i terminals, relative to other terminal types, for maintaining a constant blocking probability.

So far in this discussion, the traffic ρ_i is controlled by limiting the subscription of terminals of type i so that the probability of call blocking is bounded. In practice, many terminals exhibit multi-layer activity status, described in chapter 2 and chapter 7. For example, a communication terminal may make a subscription, which consists of a number of calls. Each call may consist of a number of bursts of information. Each burst may be segmented into a number of packets.

Consider the sharing of a transmission facility by these terminals with multi-layer traffic. Blocking of communication can occur at any one of these layers. For example, a terminal may be blocked for subscription, if allowing the subscription may cause excessive blocking of calls for terminals with completed subscription. Likewise, a call can be blocked if allowing the call completion may generate excessive blocking of bursts for terminals with completed calls. The same argument goes for blocking a burst to avoid excessive blocking (due to buffer overflow) at the packet layer. In summary, we may block terminals at layer l to control blocking at layer $l-1$.

To illustrate this multi-layer traffic control principle, we shall consider the bandwidth allocation problem for bursty calls in the next section.

8.6 Bandwidth Allocation for Bursty Calls

We assume that there is a fixed number of bursty terminals allowed for subscription. These terminals share a common bandwidth of capacity C. Furthermore, a random number N_i of type i terminals have completed the call initiation process. In parallel to the previous problem of controlling subscription, we assume that a bursty terminal of type i generates bursts of peak bit rate a_i for a mean duration b_i when the terminal is allowed to call. Furthermore, we assume the aggregate arrival of bursts for these N_i terminals is Poisson of rate γ_i, generating a traffic of $\rho_i = \gamma_i b_i$. The burst level offered traffic is

8.6 Bandwidth Allocation for Bursty Calls

denoted by W, and the burst level carried traffic is denoted by U. A burst is blocked when the offered traffic exceeds C. The bandwidth allocation problem for bursty calls is defined as how transmission capacity can be reserved per call for a prescribed maximum probability of burst blocking.

The most conservative approach is allocation according to peak bit rate. Mathematically, a call is allowed for completion if

$$C - \sum_i N_i a_i \geq 0 \tag{8.6.1}$$

For this approach, the probability of burst blocking is strictly zero. However, the absence of burst blocking is achieved only at the expense of increased probability of call blocking, which occurs when equation (8.6.1) is violated during call arrival.

The most liberal approach is allocation according to average bit rate. Mathematically, a call is allowed for completion if

$$C - \sum_i \rho_i a_i \geq 0 \tag{8.6.2}$$

By the law of large numbers, the burst level offered traffic should be close to its mean value when there are many superposed processes. Hence the probability of burst blocking diminishes in the limit that C is substantially larger than the peak bit rates a_i. In practice, it is prudent to set a positive threshold on the right hand side of equation (8.6.2) to lower the probability of burst blocking. The probability of call blocking, which occurs when equation (8.6.2) is violated, is substantially smaller than the approach of allocation according to the peak bit rate.

A third approach which uses the Chernoff bound provides a convenient link between bandwidth assignment and the probability of burst blocking. Since the mathematics for computing the probability of burst blocking is completely analogous to that for computing the probability of call blocking for the traffic engineering problem, we can use equations (8.5.1) to (8.5.7) and treat P_B as the prescribed maximum probability of burst blocking. Using the Chernoff bound, we admit a call if

$$s^*C - \mu_W(s^*) \geq \ln\frac{1}{P_B} \tag{8.6.3}$$

in which

$$\mu_W(s) = \sum_i \rho_i \hat{a}_i(s) \quad ; \quad \frac{\partial}{\partial s}\mu_W(s^*) = C \qquad (8.6.4)$$

Expressing equations (8.6.3-4) in a form for comparison with the approaches given by equations (8.6.1-2), a call is admitted if

$$C - \sum_i \rho_i \frac{\hat{a}_i(s^*)}{s^*} \geq \frac{1}{s^*}\ln\frac{1}{P_B} \qquad (8.6.5)$$

This policy is analogous to that given by equation (8.6.2), except that the peak bit rate a_i is replaced by $\hat{a}_i(s^*)/s^*$ (termed the virtual peak bit rate), and the threshold on the right hand side is a function of the prescribed maximum burst blocking probability. It can be readily shown that the virtual peak bit rate is larger than the peak bit rate for $s^* > 0$. Therefore, this measure of bandwidth assignment is a compromise between bandwidth allocations according to peak bit rate and average bit rate, and offers a convenient link to measuring blocking.

8.7 Exercises

1. We now derive equation (8.2.1) more formally.

    a. Show that the rate of call blocking is

    $$P_U(U=C)\times(K-C)\lambda = K\frac{\binom{K-1}{C}\rho^C}{\sum_{i=0}^{C}\binom{K}{i}\rho^i}\lambda$$

    b. Show that the rate of call arrival is

    $$K\frac{\sum_{i=0}^{C}\binom{K-1}{i}\rho^i}{\sum_{i=0}^{C}\binom{K}{i}\rho^i}\lambda$$

8.7 Exercises

    c. Hence show that the probability of call blocking is given by equation (8.2.1)

2. In practice, traffic is metered by sampling individual trunks in a trunk group at regular intervals (say every 30 seconds). Let us assume that the carried traffic measured per trunk is a erlang. Also, assume that there are C trunks in a trunk group, and that the offered traffic has a Poisson arrival rate.

    a. Devise a method to compute the offered traffic per trunk.

    b. What is the probability of call blocking?

3. For the Erlang blocking formula, derive the relationship:

$$B(C,\rho) = \frac{\rho B(C-1,\rho)}{C+\rho B(C-1,\rho)}$$

This recursion can be used for computing $B(C,\rho)$ accurately.

4. Very often, multi-dimensional Markov chains can be considered as one dimensional as shown in the following example. Suppose we have two types of calls with Poisson call arrival rate λ_1 and λ_2 respectively. The distributions of the active period are exponential with mean $1/\mu_1$ and $1/\mu_2$ respectively. Therefore, the offered traffic are $\rho_1 = \lambda_1/\mu_1$ and $\rho_2 = \lambda_2/\mu_2$ respectively. Show that the distribution for N, the number of calls in progress, is the same as the case when there is only 1 call type with offered traffic $\rho = \rho_1 + \rho_2$. (Hint: Let $N = N_1 + N_2$, and sum the probabilities for all N_1 and N_2 giving the same sum N.)

5. (continuation) How is the above result affected by state space truncation when the two type of calls are sharing a fixed number of trunks? Does this result depend on the assumption that the active time distributions are exponential?

6. Let us study some properties of the Erlang blocking formula.

    a. Make a rough plot of $B(C,\rho)$ versus $\gamma = \rho/C$ by considering $B(C,\rho)$ and its first derivatives at $\gamma = 0, 1,$ and ∞. Explore how the plot changes as C increases.

    b. For small values of γ and fixed C, we observe that $B(C,\rho)$ is a convex function. Let us explore the implication of this observation concerning traffic fluctuations over time. Let us take two average measurements at a trunk group over a period of time, namely the average fraction of rejected traffic $\bar{B}(C,\rho)$ and the averaged offered traffic ρ. Is the estimate $\bar{B}(C,\rho)$ an overestimate or underestimate for $B(C,\rho)$?

7. In this exercise, we develop some tight approximations for $B(C,C\gamma)$ as a function of $\gamma=\rho/C$.

   a. Apply equation (8.1.2) to show that

   $$\frac{\rho^C}{C!}e^{-\rho} \leq B(C,\rho) \leq \sum_{i=C}^{\infty} \frac{\rho^i}{i!}e^{-\rho} \stackrel{\Delta}{=} P(C,\rho)$$

   b. Apply Sterling's approximation

   $$n! \approx \sqrt{2\pi n}\, n^{n+\frac{1}{2}}e^{-n}$$

   Show that the lower bound is approximated by

   $$\frac{1}{\sqrt{2\pi C}} e^{-C[(\gamma-1)-\ln\gamma]} \leq B(C,C\gamma)$$

   c. Prove the inequality

   $$\frac{\rho^i}{i!} \leq \frac{\rho^C}{C!} \times \left(\frac{\rho}{C}\right)^{i-C}$$

   Use this inequality and the result in part b to show that the upper bound is approximated by

   $$B(C,C\gamma) \leq \frac{1}{1-\gamma} \frac{1}{\sqrt{2\pi C}} e^{-C[(\gamma-1)-\ln\gamma]}$$

   d. Now compare the upper bound derived using Sterling's approximation and the improved large deviation approximation derived in last exercise of chapter 8. (Replace y/a for the result in chapter 8 by C.) Which approximation is bigger? Optional question: Which approximation is better?

8. The insensitivity result often cannot be generalized for models of terminal more complicated than the two state (active and idle)

8.7 Exercises

model. Consider the three state (active, recuperating, and idle) model for which an active terminal goes into an intermediate recuperating state before being renewed in the idle state. Similar to the two state model, a blocked terminal is immediately renewed in the idle state again. Show why we cannot combine the active and recuperating state into a single state and then use the insensitivity result.

9. Show that the peakedness $c = v/\alpha$ for overflow traffic is approximately given by

$$c = \frac{v}{\alpha} = \frac{1}{1 - \frac{\rho}{C+1}}$$

for small α, using equation (8.4.4). Interpret this approximation.

10. Compare the probability of blocking $P_{B,i}$ for Poisson traffic with different amplitudes a_i, which was defined in equation (8.2.8).

   a. The combinatorial nature of the resource constraint $\sum_i n_i a_i \leq C$ makes it difficult to make general statements relating $P_{B,i}$ for different i. Construct examples for which $P_{B,i} > P_{B,i'}$ for the cases $a_i > a_{i'}$ and $a_i < a_{i'}$.

   b. For a_i much smaller than C, the combinatorial nature is less marked. Let us consider some approximations. Using equation (8.5.2) or the proof for the Chernoff bound, argue that

   $$P_{B,i} \approx e^{s^* a_i} P_W(W > C)$$

   Hence the ratio of the blocking probabilities is roughly given by

   $$\frac{P_{B,i}}{P_{B,i'}} \approx e^{s^*(a_i - a_{i'})}$$

   Comment on how the amplitude a_i affects blocking.

11. Make a rough plot of s^* versus C by considering the following.

   a. Evaluate s^* at $C=0$ and $C=\infty$. What value of C gives $s^* = 0$?

b. Show that s^* is a concave and monotonically increasing function of C through the following argument. From the implicit equation

$$\mu'_W(s^*) = \sum_i \rho_i a_i e^{s^* a_i} = C$$

show that

$$\frac{\partial s^*}{\partial C} = \frac{1}{\sum_i \rho_i a_i^2 e^{s^* a_i}} > 0$$

and that the second derivative of s^* with respect to C is negative.

c. Now make a better estimate of the probability of blocking

$$P_{B,i} \approx e^{s^* a_i - \frac{\partial s^*}{\partial C} a_i^2} P_W(W > C)$$

How does this new estimate compare with the estimate in the previous problem.

12. Prove the following reciprocity relationship

$$\frac{\partial P_{B,i}}{\partial \rho_{i'}} = \frac{\partial P_{B,i'}}{\partial \rho_i}$$

using equation (8.2.8) and properly normalized probabilities in equation (7.6.4).

8.8 References

Teletraffic theory and formulas of blocking were invented by the pioneer A. K. Erlang. A biography of Erlang was written by Jensen. The books by Cooper and Syski contain a more detailed treatment of teletraffic theory. The book by Kelly gives a more thorough treatment of multi-dimensional Markov chains, the truncation of state space, and the reversibility of stochastic processes, which were described briefly in this chapter.

8.8 References

Erlang was the first to suggest the insensitivity of blocking formulas to holding time distributions. Many proofs were given, such as in the papers by Burman *et. al.* and Lam, as well as in the books by Kelly and Ross. The proof in this chapter is new, which also shows that blocking is insensitive to the details of the idle time distribution.

The paper by Wilkinson contains some of the most used methods for engineering traffic in the telephone network. A more analytical treatment of blocking can be found in the paper by Jagerman. Extensions of the Erlang blocking formula were studied by Kaufman, Roberts, and Whitt. The equivalent random method was first introduced by Neal, and developed further by Wilkinson. The use of the Chernoff bound and its economic interpretation for traffic engineering and bandwidth allocation in sections 8.5 and 8.6 can be found in the article by Hui.

1. D. Y. Burman, J. P. Lehocsky, and Y. Lim, "Insensitivity of Blocking Probabilities in a Circuit-Switching Network," *J. Appl. Probability*, 21 No. 4, Dec. 1984.

2. R. B. Cooper, *Introduction to Queueing Theory*, New York: Elsevier North Holland, 1981.

3. J. Y. Hui, "A Congestion Measure for Call Admission and Bandwidth Assignment for Multi-Layer Traffic," special issue on Broadband Network Performance and Congestion Control, the *International Journal of Digital & Analog Cabled Systems*, to be published, 1990.

4. D. L. Jagerman, "Some properties of the Erlang loss functions," *Bell Syst. Tech. J*, 53, No 3, March 1974.

5. A. Jensen, "An elucidation of Erlang's statistical works through the theory of stochastic processes, in *The Life and Works of A. K. Erlang*, pp. 23-100. Trans. Danish Acad. Sci., Copenhagen, 1948.

6. J. Kaufman, "Blocking in shared resource environment," *IEEE Trans. on Commun.*, vol. COM-29, Oct. 1981.

7. F. P. Kelly, *Reversibility and Stochastic Networks*. New York: Wiley, 1979.

8. S. S. Lam, "Queueing networks with population size constraints," *IBM J. Res. Dev.*, July 1977.

9. S. R. Neal, "The equivalent group method for estimating the capacity of partial-access service systems which carry overflow traffic," *Bell Syst. Tech. J*, 51, No. 3, March 1972.

10. J. W. Roberts, "A service system with hetergeneous user requirements," *Performance of Data Communications Systems and Their Applications,* pp. 423-431, 1981.

11. S. M. Ross, *Stochastic Processes,* New York: Wiley, 1983 R. Syski, *Introduction to Congestion Theory in Telephone Systems,* Oliver and Boyd, Edinburgh, 1960.

12. W. Whitt, "Blocking when service is required from several facilities simultaneously," *AT&T Technical Journal,* vol-64, pp. 1807-1856, 1985.

13. R. I. Wilkinson, "Theories for Toll Traffic Engineering in the U.S. A.," *Bell Syst. Tech. J.*, 35, No. 2, March 1956.

CHAPTER 9

Blocking for Multi-Stage Resource Sharing

- *9.1 The Multi-Commodity Resource Sharing Problem*
- *9.2 Blocking for Unique Path Routing*
- *9.3 Alternative Path Routing- The Lee Method*
- *9.4 Assumptions for Approximations*
- *9.5 Alternative Path Routing- The Jacobaeus Method*
- *9.6 Complexity of Asymptotically Non-Blocking Networks*

The previous chapter dealt with sharing a single stage of resources such as the capacity of a transmission link. Communication in a network involves sharing multiple stages of resources. Blocking occurs when allocation fails in one of the stages. We shall compute the probability of blocking in these networks. In particular, product form solutions apply for networks with unique path routing, whereas assumptions of independence are needed for approximating blocking probability for networks with alternative path routing.

9.1 The Multi-Commodity Resource Sharing Problem

Suppose we have K processes (such as calls) sharing a set, named Υ, of resources. Each process k, $1 \leq k \leq K$, may either be in an idle state (with the activity indicator $c_k=0$) when it uses no resource from Υ, or in an active state ($c_k=1$) when it uses the resource subset $\pi_k \subset \Upsilon$. If all resources in π_k are available when the process becomes active, it seizes all the required resources, rendering them unavailable to other processes until the process returns to the idle state. The process is blocked if any resource in π_k is unavailable.

In practice, a process which has partial success in acquiring π_k may seize the available resources in π_k while waiting for the remaining resources in π_k to be available. However, we may have two or more processes holding up the resources required by each others. When none of the conflicting processes yields its acquired resources, no process can have its request satisfied. Consequently, no process can return to the idle state, and a deadlock results. A deadlock resolution algorithm is needed for deciding which process should yield the held up resources.

Take communication networks as an example. Communication calls often tie up resources before the completion of the call setup. This seizure results in overhead use of the network resources, leading to further congestion of the resources. Since congestion in turn leads to call setup failures, the repeated call setup requests may soon flood the network. This vicious cycle is often unstable by itself and may lead to a complete halt of communications in the network. Therefore, large communication networks always require some overload control algorithms for preventing such catastrophic failures and for bringing up the network if a catastrophic failure occurs. Overload control for networks is a complicated subject by itself, and therefore will not be treated here.

We broaden the original problem of multi-commodity resource sharing by considering the possibility that a request may be satisfied by more than one π_k. Let M_k be the set of π_k such that a request is satisfied if any π_k in M_k is available.

For each resource $x \in \Upsilon$, we define the Bernoulli availability random variable X with $X=1$ if x is available and $X=0$ otherwise. The system is in a blocking state for process k if the availability random variable

9.1 The Multi-Commodity Resource Sharing Problem

$$A_k = \bigvee_{\pi_k \epsilon M_k} \prod_{x \epsilon \pi_k} X = 0 \qquad (9.1.1)$$

in which $\prod$ is the usual multiplication, and $\bigvee$ is the OR sum satisfying $1 \vee 1 = 1$, $1 \vee 0 = 0 \vee 1 = 1$ and $0 \vee 0 = 0$.

Suppose a subset Ω of the K processes is active. Is there a choice of $\pi_k \epsilon M_k$ such that $A_k = 1$ for all $k \epsilon \Omega$, and $\pi_k \cap \pi_{k'} = \emptyset$, the empty set for all $k, k' \epsilon \Omega$? (The second condition ensures no single resource is seized by two different processes.) This resource assignment in general is known to be intractable, in the sense that no known assignment algorithm with a polynomial run time in K can always find such an assignment if one exists. However, certain networks have simple link assignment algorithms, such as the Clos network considered in chapter 3.

Instead of looking at the algorithmic assignment problem, we shall focus on a probabilistic view of availability. We assume a statistical description of the terminal behavior and a specified method of choosing one among the possibly many available π_k. Requests are first come first served, and resources assigned are not rearrangeable in the sense that switching to another π_k is not allowed while the process k remains active.

9.2 Blocking for Unique Path Routing

Consider first that each M_k has only 1 element, namely that there is only one way to satisfy the resource request of process k. Furthermore, let us assume that each process k is an alternating state renewal process, with mean holding time $1/\mu_k$ in the state $c_k = 1$, and mean idle time $1/\lambda_k$ in the state $c_k = 0$. We shall show that the steady state probability for the state $(c_1, \cdots, c_K)$ is a product form.

$$P_{c_1 \cdots c_K} = \eta \prod_k \left(\frac{\lambda_k}{\mu_k} \right)^{c_k} \quad \text{if } \pi_k \cap \pi_{k'} = \emptyset \text{ for all } c_k, c_{k'} = 1$$

$$P_{c_1 \cdots c_K} = 0 \qquad \text{otherwise} \qquad (9.2.1)$$

Proof

Notice the similarity between equation (9.2.1) and equation (8.3.1) of the insensitivity theorem for superposed renewal processes. Indeed, the proof for equation (9.2.1) is almost identical verbatim. The only difference is in the condition for blocking. □

Let us broaden the resource sharing problem further. Instead of being an alternating state renewal process, we may consider the process k as a superposition of many processes, such as u_k renewal processes. In the limiting case, we may also consider the process k as a Poisson process, with $u_k \leq \infty$. Therefore, each c_k has a range $0 \leq c_k \leq u_k$, instead of being 0-1. Also, let us say that the amount of the resource $x \in \Upsilon$ is given by C_x. Furthermore, the process k requires an amount $a_{kx} c_k$ of the resource $x \in \Upsilon$. The extension of (9.2.1) is straight forward:

$$P_{c_1 \cdots c_K} = \eta \prod_k \binom{u_k}{c_k} \left(\frac{\lambda_k}{\mu_k}\right)^{c_k} \qquad (9.2.2)$$

for the superposition of renewal processes, and

$$P_{c_1 \cdots c_K} = \eta \prod_k \frac{(\lambda_k/\mu_k)^{c_k}}{c_k!} \qquad (9.2.3)$$

for Poisson arrival processes. The region for which the probabilities in equations (9.2.2-3) may assume non-zero values is given by

$$\sum_k a_{kx} c_k \leq C_x \qquad \text{for all } x \in N \qquad (9.2.4)$$

The system blocking probability $P_{B,k}$ (see section 8.2) for the process k is defined as the sum of the steady state probabilities at the boundary of equation (9.2.4) for which an increment in k is prohibited.

9.3 Alternative Path Routing- The Lee Method

Before we describe the Lee method, we take a look at the problem of resource allocation for communication processes. A point-to-point communication process (labeled k in the previous discussion) in a network is defined by the two terminals involved in the communications. The resource set Υ shared by communication processes are the links in a network. A path π_k, is a set of connected

9.3 Alternative Path Routing- The Lee Method

links between two terminals. We have unique path routing if there is at most one π_k for each k, otherwise we have alternative path routing.

We are primarily interested in $P_{B,k}$ for given k. Assuming we know which terminal pair is involved, we shall drop the subscript k for all notations from this point onwards. We define the Lee graph as the part of the network containing all paths π between the terminal pair considered.

We now return to the general problem of availability for the case with multiple means of satisfying a communication request. Define the blocking random variable $B=1-A$, for which the availability random variable A was defined by equation (9.1.1). Consequently, the probability of blocking is

$$P_B = E[B] = E[1-A] = E[1- \bigvee_{\pi \in M} \prod_{x \in \pi} X] \qquad (9.3.1)$$

which is the probability of the complement of the event that at least one of the paths is available. We should realize that the expectation is not distributive over the $\bigvee$ sum. However, we can obtain an alternative expression for the probability of blocking by seeing blocking as the event that all paths are not available. In other words,

$$P_B = E[\prod_{\pi \in M}(1 - \prod_{x \in \pi} X)] \qquad (9.3.2)$$

In general, the random variables X in the same or different π are not independent. It is very difficult to account for these dependencies in computing the blocking probability. Very often, strong assumptions of independence are adopted for the sake of simplifying computation. It is important to have an intuitive grasp of how an introduced assumption affects the result. Practical path hunting algorithms often violate these assumptions. For example, path hunting is seldom random nor exhaustive: it typically has a fixed starting position, and gives up after a number of trials. Such behavior is extremely difficult to analyze.

The Lee approximation assumes that for all $x \in \Upsilon$, the associated X are mutually independent. This independence simplifies the computation for equations (9.3.1) and (9.3.2). Before proceeding to compute these equations, we make the following classifications.

1. Series-parallel networks

First, we define a parallel network, with its Lee graph defined as follows. Any two distinct routes π, π' between the input and output of the Lee graph do not share any link other than the input link and the output link as shown in figure 1. We can now generate series-parallel networks from parallel networks by replacing links in a parallel network with parallel networks as shown in figure 1. This replacement can be carried further in the resulting network. An example of parallel networks is the three-stage Clos network introduced in chapter 3. An example of series-parallel networks is the Benes network.

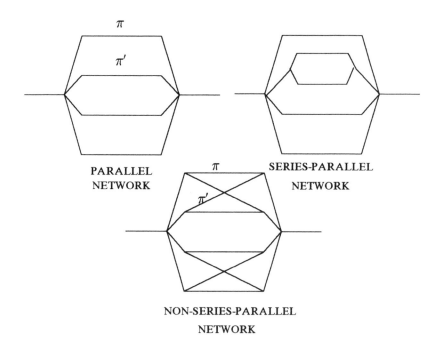

Figure 1. Series-Parallel and Non-Series-Parallel Networks

We now consider the blocking probability of parallel Lee graphs. The blocking probability of series-parallel Lee graphs can be found by solving first the blocking probabilities of the constituent parallel Lee graphs before considering that for the larger Lee graph. An example

9.3 Alternative Path Routing- The Lee Method

of this recursive solution is given in section 9.6.

Given the independence assumption in the Lee model, the blocking probability given by equation (9.3.2) becomes

$$E[B] = E[\prod_\pi (1 - \prod_{x \in \pi} X)] \tag{9.3.3a}$$

$$= \prod_\pi [1 - E[\prod_{x \in \pi} X]] \tag{9.3.3b}$$

$$= \prod_\pi [1 - \prod_{x \in \pi} E[X]] \tag{9.3.3c}$$

Equation (9.3.3b) is a consequence of the facts that the sets π are disjoint for a parallel Lee graph, and that X are independent. Equation (9.3.3c) is a consequence of the independence of X within the same path. Therefore, we can estimate $E[B]$ if we can estimate $E[X]$.

The estimation of $E[X]$ is particularly simple for multi-stage switching networks with regular structures. One such network is the rearrangeably non-blocking three-stage Clos network, which can be blocking if rearrangement is not allowed. We introduce here two symmetry assumptions for the network as well as the terminals:

- *The terminal symmetry assumption*

    The input terminals are independent and identical at the network inputs. Furthermore, an input terminal may request any idle output terminal with equal probability.

- *The network symmetry assumption*

    For multi-stage networks, the links in a stage are symmetrical. This symmetry includes a structural symmetry of the links, as well as an assumed symmetry in the choice among available routes in a parallel network.

Given these symmetry assumptions, we may assume an occupancy probability p_k for all links at the output of stage k, $1 \leq k \leq n$. Suppose there are l_k outgoing links. The parameters p_k and l_k are illustrated in figure 2 for $n=3$. For $n>3$, we assume that the paths in a Lee graph do not share edges in the following calculation. Since the number of established circuits is equal to the number of occupied links in each stage, we have the following conservation relationship:

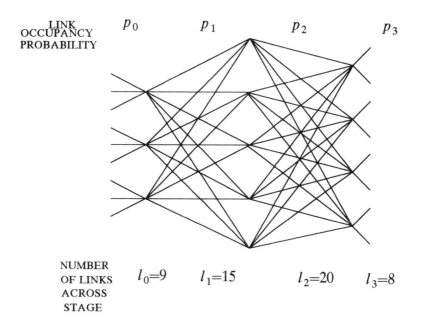

Figure 2. Occupancy Probabilities in Multi-Stage Networks

$$p_0 l_0 = p_1 l_1 = \cdots = p_k l_k = \cdots = p_n l_n \qquad (9.3.4)$$

Given p_0, we can obtain other p_k. The probability of blocking is then given by:

$$E[B] = \prod_{\pi} [\, 1 - \prod_{k=1}^{n-1}(1-p_k)\,] \qquad (9.3.5a)$$

$$= [\, 1 - \prod_{k=1}^{n-1}(1-p_k)\,]^R \qquad (9.3.5b)$$

$$= [\, 1 - \prod_{k=1}^{n-1}(1-p_0\frac{l_k}{l_0})\,]^R \qquad (9.3.5c)$$

in which R is the number of routes in a series-parallel network.

9.3 Alternative Path Routing- The Lee Method

Notice the product index ranges over $[1, n-1]$ instead of $[0, n]$ because the input link and the output link are always available to the input terminal and output terminal respectively.

It remains to find p_0. Let the offered traffic per terminal be p. The carried traffic is p_0, and the difference $p - p_0$ is the blocked traffic. Approximating $E[B]$ (a time congestion measure) by the call congestion measure, we have

$$E[B] = \frac{p - p_0}{p} \tag{9.3.6}$$

We can solve p_0 from equations (9.3.5c) and (9.3.6), consequently giving $E[B]$.

In practice, we would like to have a small blocking probability, hence $p_0 \approx p$, which can be substituted directly into equation (9.3.5c) for $E[B]$. As an example, consider the three-stage Clos network shown in figure 2. We have then:

$$E[B] \approx \left[1 - (1 - \frac{15}{9}p)(1 - \frac{20}{9}p)\right]^5 \tag{9.3.7}$$

2. Non-series-parallel networks

These networks cannot be generated by a recursive substitution of links by parallel networks. Hence equation (9.3.3b) is no longer valid because we cannot assume the independence of the availability of different π with shared links. Instead of moving the expectations inside $\prod_{x \in \pi} X$, we first expand $\prod_{\pi} [1 - \prod_{x \in \pi} X]$ into a sum of products. We then take the expectation of each term in the sum. We reduce each term in the sum by eliminating repeated X, since $X^2 = X$ for 0-1 random variables. We then take expectation of each reduced term, and apply the Lee assumption of independence for each multiplicant of a term.

This reduction process is in general difficult when the number of alternative paths is large. Very often, the reduction has to be automated by computer programs. A small example of the reduction process is given in the exercise.

9.4 Assumptions for Approximating Blocking Probabilities

Let us examine in more detail the Lee assumption that all X for $x \in \Upsilon$ are independent. We shall focus on the series-parallel Lee graph. Let us reexamine the derivation of equation (9.3.3), which was broken into two steps, namely equations (9.3.3a) and (9.3.3b). These steps constitute two kinds of independence.

- Independent path assumption - Parallel paths are available independently.
- Independent link assumption - Serial links in a path are available independently.

The Jacobaeus method improves on the Lee method by removing the independent path assumption, while keeping the independent link assumption. Before we describe the Jacobaeus method, we take a more careful look at the four assumptions (the two independence assumptions and the two symmetry assumptions of the previous section) used in computing the blocking probability. We are interested in seeing when these assumptions fail.

1. The terminal symmetry assumption

— The assumption that terminals are identical restricts us to a single service type.

— The assumption that connection requests are uniform for all idle outputs may ignore the presence of communities of interest. For example, people within a building tend to communicate more to each other than to people outside.

— The terminals are not strictly independent because terminal blockings are correlated. However, such correlation becomes weak when the blocking probability is small.

2. The network symmetry assumption

This assumption is simply a restriction on the network structure and the path hunting algorithm. With this assumption, we can view each link in a stage as being symmetrical, therefore simplifying the calculation for the link occupancy probability.

3. The independent path assumption

9.4 Assumptions for Approximating Blocking Probabilities

Consider the series-parallel Lee graph shown in figure 1. The independent path assumption assumes the link occupancies on parallel paths are independent Bernoulli random variables. However, the output links at the branching node (the node where the parallel paths π and π' diverge in figure 1) tend to have correlated occupancy. Consider a branching node with m inputs and n outputs. Furthermore, assume that the input links to the node have independent Bernoulli occupancies. Consider the following three cases, which are illustrated in figure 3.

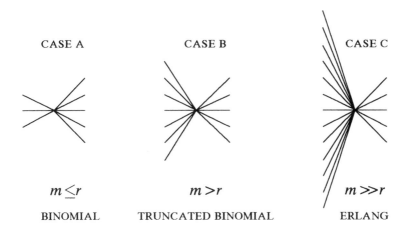

Figure 3. Distribution of Output Occupancies

— Case A: The fan-in m is smaller than the fan-out r.

Obviously, the number of occupied fan-out links, which is binomially distributed (the sum of m Bernoulli random variables) cannot be more than m. The output link occupancies have strong mutual dependence. The Lee approximation, which assumes the number of occupied fan-out links to be binomially distributed according to the sum of r Bernoulli random variables, can be inaccurate.

— Case B: The fan-in m is larger than the fan-out r.

The number of occupied fan-out links has a truncated binomial distribution, namely the binomial distribution generated by the sum of m Bernoulli random variables truncated at r.

— Case C: The fan-in m is substantially larger than the fan-out r.

In this case, we may use the Erlang distribution as a limiting approximation for the truncated binomial distribution.

For each of these three cases, the Jacobaeus approximation assumes the more accurate distribution for the number of occupied outputs than that for the Lee approximation. The Lee approximation is the same as the Jacobaeus approximation only when $m=r$. The inaccuracy of the Lee approximation can be shown by considering the strict-sense nonblocking three-stage Clos network, with $r=2m-1$. The Lee model would give a strictly positive blocking probability even though all connections can be made. As we shall show, the Jacobaeus approximation would give a strictly zero blocking probability. However, the Lee method is pretty good in practice because it produces in general a blocking probability larger than the actual value, and hence can be used for over engineering.

4. The indepei ent link assumption.

More generally, this assumption models the links in different stages of a path as independent. This approximation becomes more accurate when the fan-in and fan-out of each node are large. As a result, the contribution of a single fan-in link to a fan-out link becomes negligible.

9.5 Alternative Path Routing- The Jacobaeus Method

Having examined these four assumptions, we now describe the Jacobaeus method, which is an improvement over the Lee model by removing the independent path assumption. The Lee model assumes that the availability of each link in the network is an independent Bernoulli random variable. The Jacobaeus method improves the Lee model by assuming a more accurate description of the distribution for the number of occupied links at the input or the output of each switch node.

We shall illustrate this method using the 3-stage Clos network. The process of matching idle links is shown in figure 4. Let each link be

9.5 Alternative Path Routing- The Jacobaeus Method

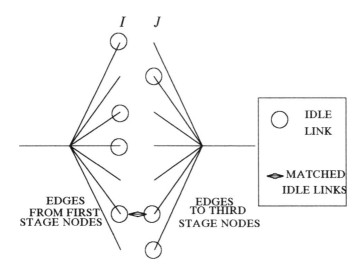

Figure 4. Matching Idle Links for 3-Stage Clos Network

named according to its top-down position. Let I be the set of idle links branching out from the first stage, and J be the set of idle links merging into the third stage. Let r be the fan-out of the first stage nodes, as well as the fan-in of the third stage nodes. Blocking occurs when $I \cap J = \emptyset$, the empty set. Hence the system blocking probability is given by

$$E[B] = \sum_{I \cap J = \emptyset} P(I,J) = \sum_{I \cap J = \emptyset} P(I) P(J) \qquad (9.5.1)$$

The product form in the last equality results from the independent link assumption. By the symmetry assumptions, two different I with the same number of idle links are equally likely. Similarly, two different J with the same number of idle links are also equally likely. Let us now count the number of different I,J combinations such that $I \cap J = \emptyset$. Let i and j be the number of non-idle branching links and non-idle merging links respectively. The event $I \cap J = \emptyset$ occurs when the set of idle merging links J (with size $r-j$) is contained in the set of non-idle branching links (with size i). This event can occur only if

$(r-j) \leq i$, or equivalently $(i+j) \geq r$. Therefore, the desired count is given by the number of combinations of i choose $r-j$, and the probability of blocking for given i, $j \leq r$ is β_{ij} given by

$$\beta_{ij} = \frac{\binom{i}{r-j}}{\binom{r}{j}} = \frac{i!j!}{(i+j-r)!r!} \quad \text{for} \quad (i+j) \geq r \quad (9.5.2)$$

$$= 0 \quad \text{otherwise}$$

Let i and j have probabilities f_i and g_j for $0 \leq i, j \leq r$. Therefore, we can express the probability of blocking of equation (9.5.1) in terms of

$$E[B] = \sum_{i,j} f_i g_j \beta_{ij} \qquad (9.5.3)$$

We shall assume either a binomial or Erlang distribution for f_i and g_j, resulting in the following 4 possible combinations:

1. Binomial f_i and binomial g_j

First, consider the case that the fan-in m is less than fan-out r for the first stage switches. The third stage switches have fan-in r and fan-out m. We assume a binomial distribution with mean mp over the range $[0,m]$ for the number of occupied fan-out (fan-in) links for the first (third) stage switches. The blocking probability is given by:

$$E[B] = \frac{(m!)^2 (2-p)^{2m-r} p^r}{r!(2m-r)!} \qquad (9.5.4)$$

Proof:

This result is obtained by a substantial amount of algebraic manipulation which we highlight in the following. As stated in equation (9.5.2), blocking can occur only if $i+j \geq r$. For a given i, blocking can occur only if j is in the range $[r-i, m]$. Since the maximum value of j is m, the minimum value of i satisfying $i+j \geq r$ is $r-m$. Hence blocking can occur only if i is in the range

9.5 Alternative Path Routing- The Jacobaeus Method

$[r-m, m]$. These ranges of summation, together with equation (9.5.3) gives:

$$E[B] = \tag{9.5.5}$$

$$\sum_{i=r-m}^{m} \sum_{j=r-i}^{m} \binom{m}{i} p^i (1-p)^{m-i} \binom{m}{j} p^j (1-p)^{m-j} \frac{i!\, j!}{(i+j-r)!\, r!}$$

By canceling and rearranging terms as well as changing the range of the inner summation, we obtain

$$E[B] = \frac{(m!)^2}{r!} \sum_{i=r-m}^{m} \frac{p^i (1-p)^{m-1}}{(m-i)!(m-(r-i))!} \times \tag{9.5.6}$$

$$\sum_{j=0}^{m-(r-i)} \frac{(m-(r-i))!}{j!(m-(r-i)-j)!} p^{j+(r-i)} (1-p)^{m-(r-i)-j}$$

The second summation resembles the sum of a binomial distribution, hence gives the value $p^{(r-i)}$. By rearranging terms and changing the range of the outer summation, we obtain

$$E[B] = \frac{(m!)^2}{r!(2m-r)!} p^r \sum_{i=0}^{2m-r} \frac{(2m-r)!}{(2m-r-i)!\, i!} (1-p)^{2m-r-i} \tag{9.5.7}$$

The binomial summation gives $(2-p)^{2m-r}$. Hence equation (9.5.7) reduces to equation (9.5.4). □

For the special case of $m=r$, the above formula gives $E[B] = (2-p)^r p^r = (1-(1-p)^2)^r$, which is just the Lee approximation. For $m \neq r$, the Lee approximation is consistently larger.

For the case of $m > r$, we use the truncated binomial distribution instead

$$f_i = g_i = \frac{\binom{m}{i} \alpha^i}{\sum_{j=0}^{r} \binom{m}{j} \alpha^j} \quad ; \quad \alpha = \frac{p}{1-p} \tag{9.5.8}$$

Unfortunately, the resulting equation for $E[B]$ cannot be readily simplified because the ranges of summations in equation (9.5.5) are changed, and hence the summations cannot be reduced.

2. *Binomial distribution for f_i and Erlang distribution for g_j*

Since the last stage can receive traffic from all inputs, we may assume an Erlang distribution for g_j, with an applied traffic of a per output. Consider first a binomial distribution for f_i, with applied traffic α per input. The probability of blocking is given by:

$$E[B] = \frac{\dfrac{a^r}{r!}}{\sum_{j=0}^{r} \dfrac{a^j}{j!}} \div \frac{\dfrac{(a/\alpha)^m}{m!}}{\sum_{j=0}^{m} \dfrac{(a/\alpha)^j}{j!}} = \frac{B(r,a)}{B(m,\dfrac{a}{\alpha})} \qquad (9.5.9)$$

which is the ratio of two Erlang distributions.

Proof:

In order to have blocking, $i+j \geq r$, hence j must range from $r-m$ to r. For a given j, i must range from $r-j$ to m. Therefore

$$E[B] = \sum_{j=r-m}^{r} \sum_{i=r-j}^{m} \binom{m}{i} \alpha^i (1-\alpha)^{m-i} \frac{a^j/j!}{\sum_k a^k/k!} \frac{i!j!}{(i+j-r)!r!} \qquad (9.5.10)$$

The method of reduction is then similar to the first case with binomial distributions at both stages. We first change the index of summation of i to range from 0 to $m-(r-j)$ and then the index of summation of i to range from 0 to m. After a substantial amount of rearrangement of terms, we obtain equation (9.5.9). □

Similarly, we may have the binomial distribution f_i truncated for $m > r$. Similar to the previous case, the expression cannot be reduced.

3. *Erlang distribution for f_i and binomial distribution for g_j*

This is completely symmetrical to the second case. Here we have a large number of terminals concentrated at the inputs. Given such heavy concentration at the inputs, the assumption that the output distribution is not Erlang makes little sense.

4. *Erlang distribution for both f_i and g_j*

9.5 Alternative Path Routing- The Jacobaeus Method

For this case, we consider heavy concentration $m \gg r$ at the input switch nodes, hence giving the Erlang distribution for the occupancy of the outputs of the input switch nodes. Consequently the distribution of the occupancy of the inputs to an output switch node can also be assumed to be Erlang. In this case,

$$E[B] = \frac{\alpha B(r,\alpha) - aB(r,a)}{\alpha - a} \qquad (9.5.11)$$

in which B is the Erlang blocking formula.

Proof:

This proof contains substantial algebraic manipulations and may be skipped.

In order for blocking to occur, i may range from 0 to r, consequently j must range from $r-i$ to r. Therefore

$$E[B] = \sum_{i=0}^{r} \sum_{j=r-i}^{r} \frac{\frac{a^i}{i!}}{\sum_{k=0}^{r} \frac{a^k}{k!}} \frac{\frac{\alpha^j}{j!}}{\sum_{k=0}^{r} \frac{\alpha^k}{k!}} \frac{i!j!}{(i+j-r)!r!} \qquad (9.5.12)$$

$$= \frac{1}{r! \sum_{k=0}^{r} \frac{a^k}{k!} \sum_{k=0}^{r} \frac{\alpha^k}{k!}} \sum_{i=0}^{r} \sum_{j=r-i}^{r} \frac{a^i \alpha^j}{(i+j-r)!}$$

Consider only the double summations. We now make successive changes for the index of summation. The first transformation (equation (9.5.13a)) changes the term $1/(i+j-r)!$ into simply $1/j!$. The second transformation (9.5.13b) exchanges the order of summation so that the term $1/j!$ may be taken out of the summation over i. The third transformation (9.5.13c) shifts the index of summation to start from $i=0$. We then sum the geometric series in the summation over i in (9.5.13d), and then reduces the expression to a symmetrical form in (9.5.13e).

$$\sum_{i=0}^{r} \sum_{j=r-i}^{r} \frac{a^i \alpha^j}{(i+j-r)!}$$

$$= \sum_{i=0}^{r} \sum_{j=0}^{i} \frac{a^{i} \alpha^{j+(r-i)}}{j!} \qquad (9.5.13a)$$

$$= \sum_{j=0}^{r} \sum_{i=j}^{r} \frac{a^{i} \alpha^{j+(r-i)}}{j!} \qquad (9.5.13b)$$

$$= \alpha^{r} \sum_{j=0}^{r} \frac{a^{j}}{j!} \sum_{i=0}^{r-j} \left(\frac{a}{\alpha}\right)^{i} \qquad (9.5.13c)$$

$$= \alpha^{r} \sum_{j=0}^{r} \frac{a^{j}}{j!} \frac{1-(a/\alpha)^{r-j+1}}{1-(a/\alpha)} \qquad (9.5.13d)$$

$$= \frac{1}{\alpha-a} \left[\alpha^{r+1} \sum_{j=0}^{r} \frac{a^{j}}{j!} - a^{r+1} \sum_{j=0}^{r} \frac{\alpha^{j}}{j!} \right] \qquad (9.5.13e)$$

Substituting equation (9.5.13e) into (9.5.12)

$$E[B] = \frac{1}{r! \sum_{k=0}^{r} \frac{a^{k}}{k!} \sum_{k=0}^{r} \frac{\alpha^{k}}{k!}} \frac{1}{\alpha-a} \left[\alpha^{r+1} \sum_{j=0}^{r} \frac{a^{j}}{j!} - a^{r+1} \sum_{j=0}^{r} \frac{\alpha^{j}}{j!} \right] \qquad (9.5.14)$$

$$= \frac{\alpha B(r,\alpha) - a B(r,a)}{\alpha - a}$$

which is equation (9.5.11). □

Of special interests, consider $\alpha \to a$. Equation (9.5.11) then gives

$$E[B] = \frac{\partial}{\partial a} \left[a B(r,a) \right] \qquad (9.5.15)$$

which can be interpreted as the rate of increase of the blocked traffic $aB(r,a)$ with respect to the offered traffic a.

9.6 Complexity of Asymptotically Non-Blocking Networks

In this section, we are interested in computing the blocking probability in networks generated by recursive applications of the

9.6 Complexity of Asymptotically Non-Blocking Networks

three-stage factorization process described in chapter 3. The resulting Lee graph is series-parallel. This recursive factorization leads to networks with complexity $O(N\log N)$, where N is the number of inputs or outputs to the network. The Benes network is an example of a network with such a complexity order.

Through a recursive application of the Lee method, we shall show that such $O(N\log N)$ networks have a blocking probability converging rapidly to a constant A which is less than 1 and greater than 0. In other words, an input-output connection can be made in this network with a certain positive probability A no matter how big N is, even if we do not allow rearrangement of circuits! Using parallel planes of such networks, we shall show that the blocking probability can be made as small as we please, while maintaining the $O(N\log N)$ complexity.

Consider the three-stage Clos network, for which the center stage is recursively factorized. The corresponding Lee graph is shown in figure 5, with r_k links coming out from each first stage node or going into each third stage node. (The index k denotes a step of the recursive factorization.) For each of these links, let $q_k = 1 - p_k$ be the probability that the link is available. Let A_k be the probability that an input-output connection can be made. Applying the Lee independence assumption, we can express A_k in terms of q_k and A_{k-1}, the probability that a connection can be made in the center stage.

$$A_k = 1 - \left(1 - q_k^2 A_{k-1}\right)^{r_k} \qquad (9.6.1)$$

Furthermore, we have the flow conservation equation

$$p_k r_k m_k = T \qquad \text{for all } k \qquad (9.6.2)$$

in which m_k is the number of switches in the first or third stage, and T is the total amount of traffic through the switching network.

This recursive relationship enables us to compute blocking probabilities for networks with $3, 5, 7, \cdots$ stages. Let us construct the $N \times N$ Benes network through the recursive application of the three stage factorization $N = r \times \frac{N}{r}$. In each step of the recursion, there are $\frac{N}{r}$ first stage (and third stage) $r \times r$ nodes, and r second stage $\frac{N}{r} \times \frac{N}{r}$ nodes. Consequently, $r_k = r$ for all k. Also, $q_k = q$ is

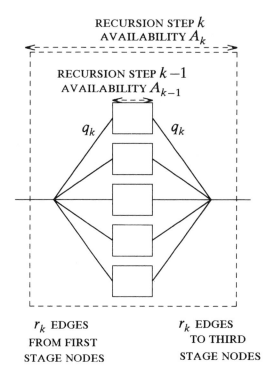

Figure 5. Lee Graph for Recursively Factorized Networks

constant for all k. Figure 6 plots equation (9.6.1) as a function of A_{k-1} for any given q and r.

Differentiating equation (9.6.1), we have

$$\frac{dA_k}{dA_{k-1}} = rq^2\left(1-q^2 A_{k-1}\right)^{r-1} \qquad (9.6.3)$$

which upon one more differentiation shows that A_k is a concave function of A_{k-1} in the interval [0,1]. Furthermore, A_k increases monotonically from 0 to a value less than 1 in this interval. Consequently, equation (9.6.1) intersects the straight line $A_k = A_{k-1}$ at the origin, and another point $0 < A_{k-1} \leq 1$ if the slope of the curve is

9.6 Complexity of Asymptotically Non-Blocking Networks

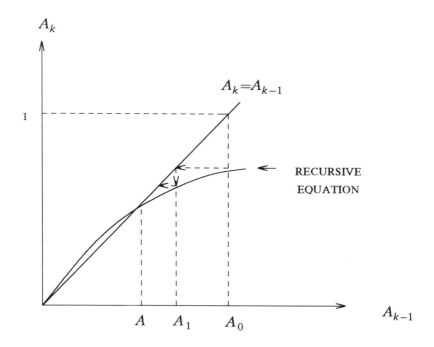

Figure 6. Recursive Computation of Availability

greater than 1 at the origin. From equation (9.6.3), this slope is greater than 1 only if

$$rq^2 > 1 \qquad (9.6.4)$$

Let us assume that this condition is satisfied. Let the second intersection be at $A_{k-1}=A_k=A>0$. Initially $A_0=1$. We can solve for A_1 from A_0 graphically as shown in figure 6. We then solve for A_2 from A_1, etc. From figure 6, A_k converges rapidly to A as a consequence of concavity of recursive equation. Consequently, the availability of connections converges to $A>0$ for large $n=\log_r N$, provided that equation (9.6.4) is valid.

We have shown that an input-output connection can be made with constant probability regardless of N and without circuit rearrangement in a Benes network. We now show that an almost non-blocking network can be constructed from these $O(N\log N)$ networks.

We use l of these networks as switching nodes in the center stage of yet another 3 stage construction. We use $l \times l$ switching nodes for the first and third stages. Therefore, there are lN inputs and lN outputs to the entire network, each with a carried traffic of q. Using the Lee model, the probability of blocking is then

$$E[B] = \left(1 - q^2 A\right)^l \tag{9.6.5}$$

Hence in order that $E[B] = \epsilon$ for given q and A, we have

$$l = \eta \log\frac{1}{\epsilon} \quad \text{where } \eta = -\frac{1}{\log(1-q^2 A)} \tag{9.6.6}$$

Let us compute the number of crosspoints required to switch a unit of traffic. The total complexity of the switching network is

$$C(r) = l \times r^2 \times \frac{N}{r} \times 2\log_r N + l^2 \times 2 \times \frac{Nl}{l} \tag{9.6.7}$$

$$= 2lrN\log_r N + 2l^2 N$$

$$= O(lN\log N) + O(lN\log\frac{1}{\epsilon})$$

For given ϵ, q, and r, the first term dominates for large N.

This network carries a total traffic of $T = lNp = lN(1-q)$. Therefore, the crosspoint complexity for switching a unit of traffic is given by

$$\frac{C(r)}{T} = \frac{2r\log_r N}{1-q} + \frac{2l}{1-q} \tag{9.6.8}$$

$$= O(\log_r N)$$

Consequently, the number of crosspoints required per unit traffic is $O(\log N)$, consistent with the combinatorial lower bound derived in chapter 3.

9.6 Complexity of Asymptotically Non-Blocking Networks

Ignoring the lower order term in equation (9.6.8), let us minimize $C(r)/T$ subject to $rq^2 > 1$. Consider $rq^2 = 1$. The minimum is obtained when $r = 5$, giving $C(r)/T = 7.8\log_2 N$.

There is a remarkable parallel between this almost non-blocking (or ϵ-blocking) construction and the Cantor network. Both employ l parallel planes of Benes networks. For the Cantor network, strict-sense non-blocking switching is achievable by having each input demultiplexed onto the $l = \log_2 N$ parallel planes. Hence, the number of crosspoints for switching a unit of traffic is $O((\log N)^2)$. The almost non-blocking network has roughly the same crosspoint count as the Cantor network, except that it serves lN inputs instead of N inputs. Hence there is an $O(\log N)$ reduction in crosspoint requirement for switching a unit of traffic for the almost non-blocking network. This gain is achieved by allowing a small blocking probability ϵ. Therefore, a significant discontinuity for crosspoint complexity occurs at $\epsilon = 0$.

9.7 Exercises

1. Draw the Lee graph for the number 5 crossbar network shown in figure 15 of chapter 4. Then, compute the probability of blocking using the Lee method.

2. For the non-series-parallel Lee graph shown in figure 7, assume that the probabilities of occupancy for each link are a, b, and c for the three stages. Apply the reduction method for non-series-parallel graph for the case of $m = n = 2$ and show that

$$E[B] = 1 - 4abc + 2ab^2c^2 + 2a^2b^2c + 2a^2b^2c^2 - 4a^2b^3c^2 + a^2b^4c^2$$

3. Using the result of problem 2, compute the probability of blocking for the non-series-parallel Lee graph of figure 1.

4. For the non-series-parallel Lee graph of figure 7, show that the probability of blocking can be approximated by

$$E[B] = \left[1 - a\left(1-(1-bc)^n\right)\right]^m$$

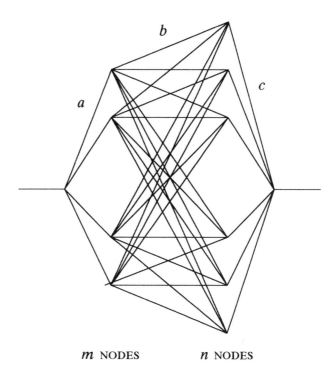

Figure 7. A Non-Series-Parallel Network

State the independence assumptions used in this approximation.

5. Consider using an $N \times N$ banyan network (with no internal buffer) for interconnecting inputs to outputs. To make a slight generalization from the banyan network made of 2×2 switches, we assume the banyan network has n stages of $d \times d$ switches, such that $N = d^n$. Each switch in a stage is connected to every switch in the next stage by a single link, and there is a unique path between each input and output of the network. Suppose each input independently requests, with probability r_0, a connection to a random (but not necessarily distinct) output.

   a. Find the probability r_k that the connection may reach stage k, $1 \leq k \leq n$ without being blocked. Assume that when more than one connection require the same output of an internal node, only one of the connection is chosen at random to further the connection.

b. Give an asymptotic approximation for r_n for large n.

6. (continuation) Repeat the above problem if the $d \times d$ switches are replaced by $ld \times ld$ switches, and there are l links connecting two switches in successive stages.

7. Compute the blocking probability according to the Lee method for a 3 stage strict-sense non-blocking 1024×1024 Clos network for which the first stage switches are of size 32×63. Assume the offered traffic per input to be $p = 0.7$. Why does the Jacobaeus method give a strictly zero blocking probability for the strict-sense non-blocking Clos network, whereas the Lee method does not?

8. Consider the three stage network with r middle stage switches. Hence there are r alternate paths. Suppose each of these paths is busy with independent probability q. We assume a random search algorithm for finding a free path. Show that the average number of search before a free path is found (or declared not found after searching all r paths) is given by $(1-q^r)/(1-q)$.

9.8 References

The work of Jacobaeus pioneered the study of blocking probability in multi-stage connecting networks. His blocking formulas also give implicitly the number of second stage connectors required to make the three-stage network strict-sense non-blocking, before this notion of non-blocking was made explicit in the seminal work and constructions of Clos (referenced in chapter 3). Variations of the Jacobaeus method are given in the paper by Karnaugh. Different notions of blocking can be found in the paper by Wolman. A study of blocking in general circuit switched networks can be found in the paper by Kelly.

The independence assumption proposed by Lee has been used extensively in evaluating the performance of switching systems. The Lee approximation usually overestimates the blocking probability. A critique of various approximation methods, including Lee's, can be found in the paper by Benes.

The paper of Ikeno applies the Lee approximation recursively and demonstrates the $O(N \log N)$ complexity for almost non-blocking networks. Section 9.6 simplified Ikeno's presentation. A more refined version of Ikeno's result can be found in the paper by Pippenger.

1. V. E. Benes, "On some proposed models for traffic in connecting networks," *Bell Syst. Tech. J.*, vol. 46, pp. 105-116, 1967.

2. N. Ikeno, "A limit on crosspoint number," *IRE Trans. Inform. Theory (Special Supplement)*, vol. 5, pp. 187-196, May 1959.

3. C. Jacobaeus, "A study of congestion in link system," *Ericsson Techniques*, 48, pp. 1-68, 1950.

4. M. Karnaugh, "Loss of point-to-point traffic in three-stage circuit switches," *IBM J. Research and Development*, vol. 18, no. 3, pp. 204-216, May 1974.

5. F. P. Kelly, "Blocking probabilities in large circuit-switched networks," *Advances in Applied Probability,*, vol 18, pp. 473-505, 1986

6. C. Y. Lee, "Analysis of switching networks," *Bell Syst. Tech. J.*, vol. 34, no. 6, Nov. 1955, pp. 1287-1315.

7. N. Pippenger, "On crossbar switching networks," *IEEE Trans. on Commun.*, vol. 23 no. 6, pp. 646-659, June 1975.

8. E. Wolman, "On definitions of congestion in communication networks," *Bell Syst. Tech. J.*, vol 44, pp. 2271-2294, 1965.

CHAPTER 10

Queueing for Single-Stage Packet Networks

- 10.1 The M/M/m Queue
- 10.2 The M/G/1 Queue - Mean Value Analysis
- 10.3 The M/G/1 Queue - Transform Method
- 10.4 Decomposing the Multi-Queue/Multi-Server System
- 10.5 HOL Effect for Packet Multiplexers
- 10.6 HOL Effect for Packet Switches
- 10.7 Load Imbalance for Single-Stage Packet Switches
- 10.8 Queueing for Multi-Cast Packet Switches

In chapter 2, we described the application of FCFS queueing for sharing network resources. This queueing mechanism differentiates packet switching from circuit switching, for which transmission resources are more rigidly scheduled. Packet switching facilitates flexible resource sharing with less scheduling complexity, at the expense of introducing random delay.

The purpose of this chapter is to derive relationships between throughput and delay for such queueing systems, given a statistical description of the interarrival time distribution and the service time distribution. We shall adopt the commonly used notation of A/B/m/C, in which A designates the interarrival time distribution (M for memoryless or exponential distribution, G for general distribution, D for deterministic duration, and *Geom* for geometric distribution with discrete sampling times, etc.), B designates the service time distribution, m denotes the number of servers, and C is the size of the waiting room for the queue.

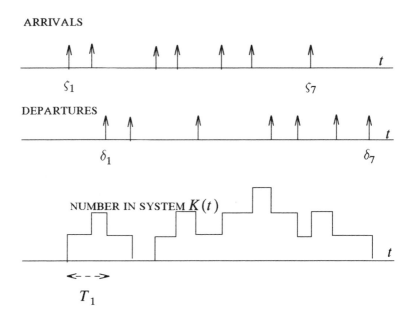

$K[i]$=0,1,2,1,0,1,2,1,2,3,2,1,2,1,0

Figure 1. The Arrival and Departure Processes

Let us describe the arrival and departure processes, which are illustrated in figure 1. The i-th arrival to the system arrives at time ς_i, and departs from the system at time δ_i. The system time for the i-th

10.1 The M/M/m Queue

arrival is given by $T_i = \delta_i - \varsigma_i$. The interarrival times $\varsigma_{i+1} - \varsigma_i$ are assumed independent, with a probability distribution described by A. An arrival at a queue with a free server would be served immediately. Otherwise, it would be served immediately when a server is freed, on the condition that all arrivals waiting in the queue are served on a first come first served basis. The service times for the arrivals are assumed independent, with a probability distribution described by B. At any time t, the number of arrivals still in the system is $K(t)$, which is equal to $i - i'$ for the largest $\varsigma_i \leq t$ and the largest $\delta_{i'} \leq t$. The discrete time sequence $K[i]$ is equal to the value of $K(t)$ right after the i-th change in the value of $K(t)$. $K[i]$ is a discrete time birth-death process since it increases or decreases by 1 at each step.

The first half of this chapter focuses mostly on the special case of the M/M/1 and the M/G/1 queues with the notation for the attribute C, assumed infinite, suppressed. These queueing models are often used because they are readily amenable to analysis. The second half considers multi-queue multi-server systems such as packet multiplexers and switches. We introduce the notation A/B/n:m/C to represent a packet system with n input queues and m servers, and C denotes the storage capacity at each input queue. We say the system is an A/B/n:m/C packet multiplexer if each arrival may be served by any one of the m available servers; and an A/B/n:m/C packet switch if each arrival specifies a server for service. For packet switches, we have to specify the probabilities of each server being requested by an arrival. Very often, we assume that the probabilities are uniform for all servers.

Multi-queue multi-server systems manifest a Head of Line (HOL) blocking phenomenon, which is a form of interference among the input queues. We described this phenomenon in chapter 6. In this chapter, we derive its statistical behavior analytically. Unlike internally non-blocking circuit switching for which we may have full loading for the inputs and the outputs, internally non-blocking packet switching is fundamentally restricted by a maximum throughput of less than 1 due to HOL blocking.

10.1 The M/M/m Queue

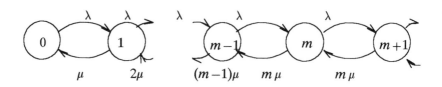

Figure 2. The M/M/m Queue with Infinite Buffer Room

Instead of clearing a blocked arrival (the focus of chapter 8), we may allow the arrival to wait in a FCFS queue until one of the m servers become available. Let us analyze the M/M/m model, assuming infinite waiting room. The associated birth-death process is shown in figure 2. Let the random variable K be the number of arrivals which remain in the system (including those being served and those in the queue) at steady state. Consider the balance of probability flux from state k to $k+1$, we have

$$p_{k+1} = \frac{\lambda}{(k+1)\mu} p_k = \frac{a^{k+1}}{(k+1)!} p_0 \qquad \text{for } 0 \leq k \leq m-1 \quad (10.1.1)$$

$$p_{k+1} = \frac{\lambda}{m\mu} p_k = \frac{a^m}{m!} \left(\frac{a}{r}\right)^{k-m} p_0 \qquad \text{for } k \geq m$$

in which $a = \frac{\lambda}{\mu}$. By normalizing the probabilities, we have

$$p_0 = \left[1 + a + \frac{a^2}{2!} + \cdots + \frac{a^m}{m!}\left(1 + \frac{a}{m-a}\right)\right]^{-1} \quad (10.1.2)$$

In particular, the probability that an arrival has to wait in queue is given by

$$\sum_{k=m}^{\infty} p_k = \frac{\dfrac{m}{m-a}\dfrac{a^m}{m!}}{1 + a + \dfrac{a^2}{2!} + \cdots + \dfrac{a^m}{m!}\left(1 + \dfrac{a}{m-a}\right)} \quad (10.1.3)$$

10.1 The M/M/m Queue

This expression is known as the Erlang delay formula.

The M/M/1 queue is a special case for which we have

$$p_k = (1-a)a^k \tag{10.1.4}$$

for all $k \geq 0$. The mean number of processes in the system is

$$E[K] = \sum_{k=0}^{\infty} k p_k = \frac{a}{1-a} = \frac{\lambda}{\mu - \lambda} \tag{10.1.5}$$

Let T be the time a particular process spends in the system, including both the queueing time and the service time. Suppose there are k processes in the system when the process arrives. Since the service process is memoryless, it will still take an average time of $1/\mu$ to complete the service for the process at the head of the line. For the remaining k processes (including the process considered), the average total service time is k/μ. Hence

$$E[T \mid k] = (k+1)\frac{1}{\mu} \tag{10.1.6}$$

Consequently

$$E[T] = \sum_{k=0}^{\infty}(k+1)\frac{1}{\mu}p_k = \frac{E[K]+1}{\mu} = \frac{\frac{\lambda}{\mu-\lambda}+1}{\mu} = \frac{1}{\mu-\lambda} \tag{10.1.7}$$

Comparing equation (10.1.5) and (10.1.7), we have

$$E[K] = \lambda E[T] \tag{10.1.8}$$

which is known as Little's result. Little's result is valid under very general situations, such as general and dependent inter-arrival and service time distributions. We shall only outline an intuitive argument for this result: Given that the system can reach equilibrium, a process should find the same average number of processes $E[K]$ in the system at the times of its arrival and departure. In the period T between these two events, the average total number of arrivals is given by $\lambda E[T]$. This is simply the number of processes left behind in the system, namely $E[K]$.

10.2 The M/G/1 Queue - Mean Value Analysis

We showed in chapter 8 that the steady state distribution of the number of processes in a system with no waiting room depends only on the mean service time. Unfortunately, this insensitivity no longer holds when queueing is allowed.

We consider M/G/1 queues and speak in terms of service via transmission of variable length packets with general distribution of packet length. We assume that packets arrive independently. Thus, we may assume a Poisson arrival rate λ. We assume also that the random lengths X of packets are independent, and the probability density function for $X=x$ is given by $b(x)$.

In the continuum of time, we want to seek certain points in time for which we have a concise description of the state of the system. In essence, we look for instants of renewal. The states at these instants constitute an embedded discrete time Markov process. We used this embedding approach to analyze the semi-Markov process in Chapter 7. For the M/G/1 queue, the instants when a packet finishes being transmitted constitute such points as shown in figure 3. Let $K^-[i]$ be the number of packets left behind in the queue upon the departure of the i-th packet. The state of the queueing system at such departure times is fully represented by the discrete time random process $K^-[i]$, since none of these packets has been served at all. $K^-[i]$ can be derived very simply from $K[i]$ shown in figure 1: by deleting from the sequence any $K[i]$ which is equal to $K[i-1]+1$. For example, $K^-[i]=0,1,0,1,2,1,1,0, \cdots$ from figure 1.

An immediate question then is how the number of packets left behind upon a departure is related to the number of packets in the queue at other instants. The answer, fortunately, is that the probability distributions for both numbers are identical in steady state, as shown by the following argument.

Let us also define the sequence $K^+[i]$ as the number of packets in the system just before the i-th packet arrival. Similar to $K^-[i]$, $K^+[i]$ is obtained from $K[i]$ by deleting those $K[i]$ satisfying $K[i]=K[i+1]-1$. For example, $K^+[i]=0,1,0,1,1,2,1,0, \cdots$ from figure 1. Therefore, $K^+[i]$ and $K^-[i]$ sample the continuous time process $K(t)$ right before an arrival and right after a departure respectively.

Consider now $K^+[i]$ and $K^-[i]$ in steady state. We now show that the steady state random variables K^+ and K^- have the same

10.2 The M/G/1 Queue - Mean Value Analysis

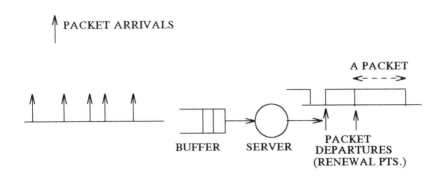

Figure 3. Renewal At Departure Time for M/G/1 Queue

probability distribution. This follows from the fact that $K[i]$ is a birth-death process which changes by +1 or -1 each time. In steady state, we can match each +1 (a rising edge from k to $k+1$ in figure 1) by a later -1 (a falling edge from $k+1$ to k in figure 1). Hence the fraction of k in the steady state sequence $K^+[i]$ is equal to the fraction of k in the steady state sequence $K^-[i]$. (For the sequences $K^+[i]$ and $K^-[i]$ in figure 1, they both have 4 ones, 1 two, and 3 zeros.)

Given this fact, we proceed to show that both K^+ and K^- have the same distribution as that for the sampled value of $K(t)$ in steady state. Since $K^+[i]$ samples $K(t)$ right before an arrival, and since arrivals are independent of the state of the system, the arrivals sample the queue at random time.

Having shown that $K^-[i]$ in steady state also describes the queue length at random time, we now proceed to compute the mean $E[K^-]$. Let us examine the state transition probabilities for $K^-[i]$. Between 2 consecutive departures, suppose we have α arrivals to the queue. For notational simplicity, let $K'=K^-[i+1]$ and $K=K^-[i]$. We can readily write down the following dynamic equation.

$$K' = K - \epsilon(K) + \alpha \tag{10.2.1}$$

where $\quad \epsilon(K) = \begin{array}{l} 0 \text{ if } K=0 \\ 1 \text{ if } K>0 \end{array}$

Before we study in more detail the above dynamic equation, we derive first the statistics of the number of arrivals α in between renewals. In steady state, the random variable α depends on the packet arrival rate λ and the transmission time for the current packet. Hence, the probability that m packets arrive while the current packet is transmitted is given by

$$P_\alpha(m) = \int_0^\infty \frac{(\lambda x)^m}{m!} e^{-\lambda x} b(x) dx \qquad (10.2.2)$$

For slotted fixed length packets, we may assume α to be a 0-1 random variable with $E[\alpha]=\lambda$. The packet interarrival time therefore has a geometric distribution, namely the probability of having to wait i slots before a packet arrives is given by $\lambda(1-\lambda)^{i-1}$. If we consider many slotted links being multiplexed onto 1 link, we may model the number of packets arriving per slot time waiting to be multiplexed as being Poisson distributed, hence

$$P_\alpha(m) = \frac{\lambda^m}{m!} e^{-\lambda} \qquad (10.2.3)$$

where λ is the expected number of packet arrival per slot.

With a statistical description of α, we are ready to examine the mean values of equation (10.2.1) through taking expectations.

$$E[K'] = E[K] - E[\epsilon(K)] + E[\alpha] \qquad (10.2.4)$$

In steady state, we have $E[K']=E[K]$. Therefore

$$E[\alpha] = E[\epsilon(K)] \qquad (10.2.5)$$

Define

$$\beta = E[\epsilon(K)] = 0 \cdot p_{k=0} + \sum_{k=1}^\infty 1 \cdot p_k = P(K>0)$$

Hence, β is the probability of finding the system busy. Equation (10.2.5) states that the expected number of arrivals between renewals is equal to the probability β of finding the system busy.

In order to find the mean queue length $E[K]$, we square both sides of the dynamic equation (10.2.1):

10.2 The M/G/1 Queue - Mean Value Analysis

$$K'^2 = K^2 + \epsilon(K) + \alpha^2 - 2K + 2\alpha K - 2\epsilon(K)\alpha \qquad (10.2.6)$$

in which we have made use of the fact that $\epsilon^2 = \epsilon$ and $\epsilon(K)K = K$ because ϵ is a 0-1 random variable. We now take expectations for both sides of equation (10.2.6). Given independence of the terms in each product, and using the equation $E[K'^2] = E[K^2]$ in steady state, we have

$$E[\epsilon(K)] + E[\alpha^2] - 2E[K] + 2E[\alpha]E[K] - 2E[\epsilon(K)]E[\alpha] = 0 \qquad (10.2.7)$$

Substituting our previously derived equation $E[\epsilon(K)] = E[\alpha]$, and after rearranging terms, we obtain the following mean-value formula:

$$E[K] = E[\alpha] + \frac{E[\alpha(\alpha-1)]}{2(1-E[\alpha])} \qquad (10.2.8)$$

For given λ and $b(x)$, we can obtain $E[\alpha]$ and $E[\alpha(\alpha-1)]$ using equation (10.2.2).

$$E[\alpha] = \sum_{m=0}^{\infty} m P_\alpha(m) \qquad (10.2.9)$$

$$= \int_0^\infty \left\{ \sum_{m=0}^{\infty} m \frac{(\lambda x)^m}{m!} e^{-\lambda x} \right\} b(x) dx$$

$$= \int_0^\infty \lambda x b(x) dx$$

$$= \lambda E[X]$$

Similarly, we have

$$E[\alpha(\alpha-1)] = \sum_{m=0}^{\infty} m(m-1) P_\alpha(m) \qquad (10.2.10)$$

$$= \int_0^\infty \left\{ \sum_{m=0}^{\infty} m(m-1) \frac{(\lambda x)^m}{m!} e^{-\lambda x} \right\} b(x) dx$$

$$= \int_0^\infty \lambda^2 x^2 b(x) dx$$

$$= \lambda^2 E[X^2]$$

Substituting equations (10.2.9-10) into equation (10.2.8), we obtain what is known as the Pollaczek-Khinchin mean-value formula.

$$E[K] = \lambda E[X] + \frac{E[X^2]}{2(1-\lambda E[X])} \qquad (10.2.11)$$

This formula will be derived in the exercises using a simpler method. We used the above method because the dynamic equation (10.2.1) for K also gives the probabilities for K, instead of just the mean value. This will be explored in the next section.

10.3 The M/G/1 Queue - Transform Method

In order to compute higher moments of the queue length K or the steady state probabilities p_k, we study the z transform of K defined by

$$\mathbf{K}[z] = E[z^K] = \sum_{k=0}^{\infty} p_k z^k \qquad (10.3.1)$$

which is related to the moment generating function $\psi_K(s) = E[e^{sK}]$ by the substitution $z = e^s$.

Instead of using the state dynamic equation (10.2.1), we broaden the scope by assuming that the HOL packet is served with probability q, which is independent of all past history of packet service or the state of the queue. Therefore, the state dynamic equation becomes

$$K' = K - \gamma \epsilon(K) + \alpha \qquad (10.3.2)$$

in which γ is a 0-1 random variable with $E[\gamma] = q$. Transforming the state equation, we have

$$\mathbf{K}[z] = E[z^K] = E[z^{K-\gamma\epsilon(K)+\alpha}] \qquad (10.3.3)$$

$$= E[z^\alpha] E[z^{K-\gamma\epsilon(K)}]$$

$$= E[z^\alpha]\left(p_0 + E[\sum_{k=1}^{\infty} p_k z^{k-\gamma}]\right)$$

10.3 The M/G/1 Queue - Transform Method

$$= E[z^\alpha]\left(p_0(1-E[z^{-\gamma}]) + E[\sum_{k=0}^{\infty} p_k z^{k-\gamma}]\right)$$

$$= E[z^\alpha]\left(p_0(1-E[z^{-\gamma}]) + \mathbf{K}[z]E[z^{-\gamma}]\right)$$

Solving for $\mathbf{K}[z]$ in the above equation, we have

$$\mathbf{K}[z] = p_0 E[z^\alpha] \frac{1-E[z^{-\gamma}]}{1-E[z^{-\gamma}]E[z^\alpha]} \tag{10.3.4}$$

In the remainder of this section, we shall evaluate equation (10.3.4) for the special cases of M/G/1 queues and *Geom/Geom*/1 queues.

1. Variable length packets and the HOL packet is always served.

This is simply the M/G/1 queue. The HOL packet is always served implies $\gamma=1$ with probability 1. Hence $E[z^{-\gamma}]=z^{-1}$. Each packet has variable length according to the density function $b(x)$. Furthermore, the number of arrivals during the service for the HOL packet is α with probabilities given by equation (10.2.2). Consequently, the transform of α is given by:

$$E[z^\alpha] = \sum_{m=0}^{\infty} P_\alpha(m) z^m \tag{10.3.5}$$

$$= \int_0^\infty \left(\sum_{m=0}^{\infty} \frac{(\lambda x)^m}{m!} e^{-\lambda x} z^m\right) b(x) dx$$

$$= \int_0^\infty e^{-\lambda x(1-z)} b(x) dx$$

$$= \psi_B(-\lambda(1-z)) \quad \text{where} \quad \psi_B(s) = \int_0^\infty e^{sx} b(x) dx$$

is the moment generating function for $b(x)$.

Hence equation (10.3.4) becomes what is known as the Pollaczek-Khinchin transform equation:

$$\mathbf{K}[z] = p_0 \frac{(z-1)\psi_B(-\lambda(1-z))}{z-\psi_B(-\lambda(1-z))} \qquad (10.3.6a)$$

$$= (1-\lambda E[X])\frac{(z-1)\psi_B(-\lambda(1-z))}{z-\psi_B(-\lambda(1-z))} \qquad (10.3.6b)$$

There are two ways to show that $p_0 = 1 - \lambda E[X]$ in which $E[X] = \int xb(x)dx$. The first way relates p_0, which is equal to the probability that the server is idle, to the amount of work done by the server in steady state. We leave the working out of this intuitive argument as an exercise for the reader. The second way sets $\mathbf{K}[1]=1$ in equation (10.3.6a). The numerator and denominator of the right hand side of equation (10.3.6a) both become zero as $z \to 1$. To evaluate this indeterminate form, we apply L'Hospital's rule by differentiating separately the numerator and the denominator with respect to z before passing the limit $z \to 1$. The denominator then becomes $1 - \lambda \psi_B'(0) = 1 - \lambda E[X]$. The numerator then becomes $\psi_B(0) = 1$. Consequently, $\mathbf{K}[1]=1$ implies $p_0 = 1 - \lambda E[X]$.

The mean of K can be easily found by differentiating $\mathbf{K}[z]$ with respect to z and then evaluating the derivative at $z = 1$. In the process, L'Hospital's rule also has to be applied. As expected, we obtain the Pollaczek-Khinchin mean-value formula. We leave the derivation to the reader.

2. Slotted Bernoulli arrivals and Bernoulli HOL service.

We are going to derive the steady state probabilities of K for fixed length packets with Bernoulli arrivals. Therefore, $\alpha=1$ with probability λ. The HOL packet is served ($\gamma=1$) with probability q in each slot. Both the interarrival time and the service time distributions are geometric. Consequently, the queue is $Geom/Geom/1$.

The transforms for γ and α are

$$E[z^{-\gamma}] = (1-q) + qz^{-1} \qquad (10.3.7a)$$

$$E[z^{\alpha}] = (1-\lambda) + \lambda z \qquad (10.3.7b)$$

Substituting equations (10.3.7a,b) into equation (10.3.4), we have

$$\mathbf{K}[z] = p_0 \left[(1-\lambda)+\lambda z\right] \frac{q(1-z^{-1})}{1 - \left[(1-q)+qz^{-1}\right]\left[(1-\lambda)+\lambda z\right]} \qquad (10.3.8)$$

10.3 The M/G/1 Queue - Transform Method

The denominator has a root at $z=1$. Thus the factors $z-1$ in both the numerator and the denominator cancel out each other, giving

$$\mathbf{K}[z] = p_0 \frac{1-\lambda+\lambda z}{(1-\lambda)q - \lambda(1-q)z} \tag{10.3.9}$$

$$= c \frac{1-\lambda+\lambda z}{1-\omega z} \quad \text{with} \quad \omega = \frac{\lambda(1-q)}{(1-\lambda)q} \quad \text{and} \quad c = \frac{p_0}{(1-\lambda)q}$$

The constant c can be found be setting $\mathbf{K}[1]=1$, hence obtaining $c=1-\omega$. To obtain the individual steady state probabilities p_k, we expand the z transform into a series

$$\mathbf{K}[z] = \sum_{k=0}^{\infty} p_k z^k \tag{10.3.10}$$

$$= (1-\omega)\frac{1-\lambda+\lambda z}{1-\omega z}$$

$$= (1-\omega)(1-\lambda+\lambda z)\sum_{k=0}^{\infty} \omega^k z^k$$

Consequently, we have

$$p_0 = (1-\omega)(1-\lambda) = 1 - \frac{\lambda}{q} \tag{10.3.11}$$

$$p_k = (1-\omega)\left[(1-\lambda)\omega + \lambda\right]\omega^{k-1} \tag{10.3.12}$$

$$= \left(1-\frac{\lambda}{q}\right)\frac{\omega^k}{1-q} \quad \text{for } k>0$$

The expected value of K can be obtained after some algebraic manipulations.

$$E[K] = \frac{\lambda(1-\lambda)}{q-\lambda} \tag{10.3.13}$$

The average delay T is given by Little's result:

$$E[T] = \frac{E[K]}{\lambda} = \frac{1-\lambda}{q-\lambda} \tag{10.3.14}$$

Suppose the input queue has a finite buffer size for S packets. The probability of buffer overflow is bounded by the probability of $K>S$ for the case of infinite buffer size. Therefore

$$P(\text{Packet Loss}) \leq P(K>S) = \sum_{k=S+1}^{\infty} \left(1-\frac{\lambda}{q}\right)\frac{\omega^k}{1-q} = \frac{\lambda}{q}\omega^S \quad (10.3.15)$$

This completes our discussion on the *Geom/Geom/1* queue, which shall be used to model input queueing for slotted packet switches in the following sections.

10.4 Decomposing the Multi-Queue/Multi-Server System

From this point onwards, we shall be mostly concerned with fixed length packet systems, either with Bernoulli or Poisson arrivals per slot. Some of the following results can be extended for exponentially distributed packet lengths as shown in the exercises. Furthermore, we assume an infinite buffer size for the input queues unless stated otherwise.

Consider the case of many FCFS queues served by a number of servers as shown in figure 4. Each server may transmit one packet from one of the input queues per slot time. We may assume a crossbar switch between the input queues and the servers.

For packet switching, a packet specifies an output for delivery. For packet multiplexing, a packet can use any output for delivery. Conflicting requests for an output are chosen in random for delivery for both cases.

Since we impose the FCFS condition, it is possible that some servers are idle even though there are packets behind the HOL packet. This phenomenon is known as HOL blocking, which occurs for both packet multiplexing and packet switching. For packet multiplexing, we may have idle servers when there are fewer non-empty queues than servers, even though each queue may have more than one packet waiting for service. For packet switching, the HOL packet can be blocked by other HOL packets destined for the same output, while a packet behind the HOL packet could have been served by a free server if the HOL packet could be preempted.

For packet multiplexers with N inputs and L outputs, we may characterize the entire system as a discrete time Markov process with

10.4 Decomposing the Multi-Queue/Multi-Server System

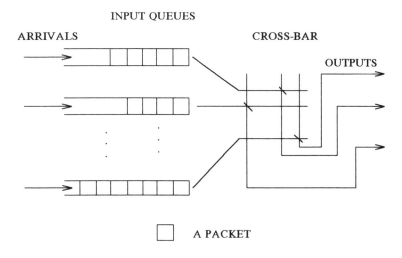

Figure 4. Multi-Queue Multi-Server System

a state space represented by the vector $(K_1, K_2, \cdots, K_i, \cdots, K_N)$, in which the random variable K_i is the number of packets in the input queue i. This characterization is illustrated in figure 5. The number of arrivals per slot may either be Poisson or Bernoulli with parameter λ_i for input i. Let the random variable H be the number of inputs with $K_i > 0$, namely, H is the number of inputs with a HOL packet. All HOL packets are served if $H \leq L$. Otherwise, we randomly choose any L out of the H HOL packets for service. The remaining $H-L$ HOL packets are blocked.

For packet switches, the state space is far more complicated. Since each packet also carries a destination address, the complete characterization of each state would include the destination addresses of the packets in the queues. The state space could be simplified if we assume that the packet addresses are independent for all packets. In this case, we may ignore the destination addresses of packets not at HOL since they are independent. However, the packet addresses for the HOL packets which have been blocked before are not independent. There is a tendency that packets with the same destination accumulate at the HOL positions. Therefore, we have to

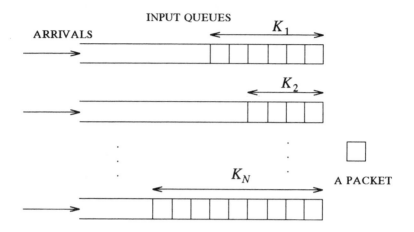

COMPLETE CHARACTERIZATION OF MARKOV PROCESS :

1. MULTIPLEXER - $K_1, \cdots, K_N$

2. PACKET SWITCH - $K_1, \cdots, K_N$ + HOL ADDRESSES

Figure 5. States for Packet Multiplexers and Switches

specify the destination addresses of the HOL packets in addition to the values K_i to obtain a complete characterization of the state space.

For simplicity of analysis, we assume that $\lambda_i = \lambda$ for all i. For packet switches, we assume that the output addresses are equally likely for all outputs. The first assumption is called the balanced input load assumption. The second assumption is called the balanced output load assumption. We shall argue towards the end of this chapter that load imbalance can be handled by a proper speed-up of the switching network. This speed-up factor depends on the maximum throughput for the balanced load case.

For both packet multiplexers and switches, an exact solution of this multi-dimensional Markov chain becomes intractable as N increases. For the sake of engineering approximations, certain

10.4 Decomposing the Multi-Queue/Multi-Server System

independence assumptions are made so that the state space can be decomposed into smaller and loosely coupled subspaces. These assumptions have been justified by simulation. This decomposition is shown in figure 6 and explained as follows.

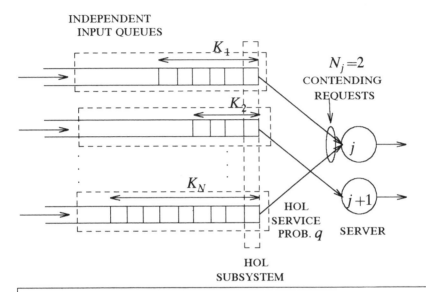

Figure 6. Decomposing a Multi-Queue Multi-Server System

Generally speaking, we treat each queue separately as the input queueing systems. The HOL locations are considered as a separate system called the HOL system. We assume that each HOL packet is served with the same probability q independent of the past history of packet arrivals and services.

Each input queueing system is a single-queue single-server system, in which the server may provide service with probability $q = E[\gamma]$. Therefore, the dynamics of the number of packets in the queue is governed by the state dynamics equation (10.3.2), and the analysis of the previous section for the *Geom/Geom/*1 queue applies for each input queue.

The value of q should take the HOL blocking phenomenon into account. The question then is how we may relate q to the degree of HOL blocking, as well as the arrival rates λ. This relationship can only be found by examining the dynamics of the HOL system.

The state of the HOL system is described by the occupancy (plus addresses for packet switches) of the HOL positions. For packet multiplexing, it suffices to consider the number of HOL packets present. For packet switching, we also have to specify the multiplicity N_j of each output address j for these HOL packets.

There are two methods to analyze the HOL system modeled as such, one approximate and the other exact. The approximate method assumes that the occupancy (as well as the address for packet switches) are independent for all HOL positions. The exact method takes the dependence into account. We shall describe both methods in the following sections 10.5 and 10.6.

Before giving an account of the dynamics of the HOL system, we first derive flow conservation relationships for the input queueing system and the HOL system.

1. Flow conservation in input queueing system

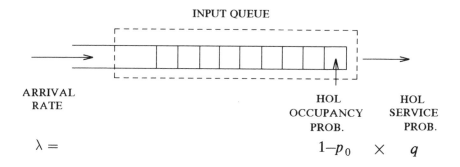

Figure 7. Traffic Flow for Input Queues

10.4 Decomposing the Multi-Queue/Multi-Server System

The flow into and out of an input queue is shown in figure 7. Notice that a packet entering the entire packet switching system first arrives at an input queue, and then moves to the HOL position before it is served. Let p_0 be the probability that a queue is empty during a slot time. Thus $1-p_0$ is the probability that there is a HOL packet. According to our independence assumption concerning service for the HOL packet, this packet is served with probability q. Thus $(1-p_0)q$ is the average number of packets served at the HOL per slot. In steady state, this must be equal to the influx λ of packets to the queue. Hence we have the following packet conservation relationship:

$$(1-p_0)q = \lambda \qquad (10.4.1)$$

We have to derive another relationship between q and p_0 in order to solve these two values for given λ. This relationship is given by considering the state transition dynamics of the HOL system.

2. Flow conservation through HOL system

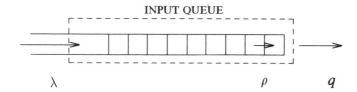

λ : ARRIVAL RATE
ρ : ARRIVAL RATE INTO HOL POSITION IF UNBLOCKED
q : DEPARTURE RATE GIVEN HOL POSITION OCCUPIED

Figure 8. Traffic Flow Through Each HOL Position

Packets can only enter into the HOL system in queues with the HOL positions not blocked in the previous slot. Let M be the random variable for the number of such unblocked HOL positions. For any one of these M positions, let ρ be the probability that a packet moves into the HOL position in the current slot as shown in figure 8. In steady state, flow conservation implies that

$$E[M]\rho = N\lambda \tag{10.4.2}$$

which states that the total arrival rate must be equal to the flow into the $E[M]$ unblocked HOL positions. We introduce M and ρ here for the purpose of parameterizing the arrival process into the HOL system. The parameter ρ characterizes the arrival rate into the HOL system, whereas the service rate q characterizes the departure rate from the HOL system.

10.5 HOL Effect for Packet Multiplexers

We now analyze the queueing behavior for a packet multiplexer with N input queues and L outputs. We assume independent and Bernoulli arrivals for each input.

As we mentioned before, there are two approaches for computing the HOL effect. The first approach assumes independence of HOL occupancies, while the second approach takes the dependence into account. Unfortunately, the second approach is rather difficult for packet multiplexers.

Let us assume that there is a HOL packet at a tagged input queue. Let H be the random number of HOL packets. From the point of view of the tagged input queue, there are $H-1$ HOL packets at the other $N-1$ HOL positions. Let π_h be the steady state probabilities for H. Ignoring the dependency of the HOL occupancies, we assume that each HOL position is occupied with an independent probability $1-p_0$. Hence, $H-1$ has a binomial distribution, namely,

$$\pi_h = \binom{N-1}{h-1}(1-p_0)^{h-1}p_0^{N-h} \quad \text{for } 1 \leq h \leq N \tag{10.5.1}$$

We now compute q, the probability that a HOL packet is served. If $h \leq L$, the packet will be served. Otherwise the packet will be served with probability L/h, assuming random choice of L packets out of the h HOL packets. Thus the probability q is obtained by averaging the chances of being served over h from 1 to N. In other words:

$$q = \sum_{h=1}^{L} 1 \cdot \pi_h + \sum_{h=L+1}^{N} \frac{L}{h} \cdot \pi_h \tag{10.5.2}$$

10.5 HOL Effect for Packet Multiplexers

Since the π_h are functions of p_0, equation (10.5.2) expresses q as a function of p_0.

From the flow conservation equation (10.4.1) for the input queues, we obtain another expression for q in terms of p_0:

$$q = \frac{\lambda}{1-p_0} \qquad (10.5.3)$$

Plotting q in equations (10.5.2) and (10.5.3) versus p_0, the cross over point for these two curves gives the unique solution for q and p_0. Given q and λ, the service probability and arrival probability to the input *Geom/Geom*/1 queue, the input queueing behavior is now completely characterized.

10.6 HOL Effect for Packet Switches

We now analyze the queueing behavior for a packet switch with N inputs and N outputs. We shall focus on the dynamics at the HOL positions. By analyzing the dynamics and the flow conservation relationships, we can derive the probability q that a HOL packet is served in a slot. This in turn will give the throughput and delay behavior of a packet switch.

There are two types of HOL packets at an input. The first kind is the blocked HOL packets that lost the contention in the previous slot. The second kind is the fresh HOL packets, which just arrive and make a first request for service by an output. The fresh HOL packets have independent destination addresses, whereas the addresses of the blocked HOL packets are not independent because they were involved in the previous contention for an output.

We discuss first the analysis which assumes independence of HOL occupancies as well as HOL addresses. This independence can be realized by changing the destination address of a HOL packet randomly whenever it is blocked. We would like to see how the HOL packet service probability q is related to the arrival rate λ.

Let H be the random number of HOL packets. The probability that none of these H packets is destined for a particular output is $(1-1/N)^H$. Therefore, the probability that one or more than one of these H packets is destined for a particular output is $1-(1-1/N)^H$. This probability also gives the throughput for the output, since a packet is delivered to the output when one or more of the packets is destined for that output. Taking the expectation over H, the expected throughput for an output is given by

$$1-E[(1-\frac{1}{N})^H] = 1-\sum_{h=0}^{N} (1-\frac{1}{N})^h \binom{N}{h}(1-p_0)^h p_0^{N-h} \quad (10.6.1)$$

$$= 1-\left[p_0+(1-p_0)(1-\frac{1}{N})\right]^N$$

$$= 1-\left[1-\frac{1}{N}(1-p_0)\right]^N$$

$$\approx 1-e^{-(1-p_0)}$$

In steady state, this must be equal to λ, the packet arrival rate. This throughput is maximized at saturation, namely at $p_0 = 0$, giving a maximum throughput of $\lambda = 1-e^{-1} = 0.632$.

Below saturation, setting equation (10.6.1) equal to λ gives

$$p_0 = 1+\log_e(1-\lambda) \quad (10.6.2)$$

The rate of service q at the HOL is given by the conservation of flow equation $(1-p_0)q = \lambda$. Knowing q and λ, we can calculate the queueing delay for the *Geom/Geom/*1 input queues.

We now proceed to consider the more realistic case with correlation of output addresses for the blocked HOL packets. Let us examine several relationships for quantities of interest in this system.

1. Dynamic Equation for the number of unblocked HOL positions M

Let us examine the dynamics of the HOL system as shown in figure 9. The number of unblocked inputs M in the next slot time is equal to the total number of inputs N, minus the total number of inputs which remain blocked in the current slot. Suppose there are N_j HOL packets with destination j in the current slot. The number of packets delivered to j during the current time slot is $\epsilon(N_j) = \min(1, N_j)$. The other $N_j - \epsilon(N_j)$ packets for destination j remain blocked. Hence

10.6 HOL Effect for Packet Switches

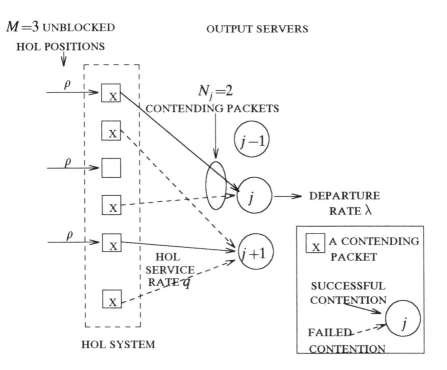

Figure 9. Traffic Flow in HOL System

$$M = N - \sum_{j=1}^{N} [N_j - \epsilon(N_j)] \tag{10.6.3}$$

We now take expectations of this equation. Dividing throughout by N and making use of the symmetry of j, we have

$$\frac{E[M]}{N} = 1 - \left(E[N_j] - E[\epsilon(N_j)]\right) \tag{10.6.4}$$

Realizing that the throughput for destination j, namely $E[\epsilon(N_j)]$, must be equal to the arrival rate λ in steady state, equation (10.6.4) becomes

$$\frac{E[M]}{N} = 1 - \left(E[N_j] - \lambda\right) \tag{10.6.5}$$

Using the flow conservation relationship $E[M]\rho = N\lambda$ given in equation (10.4.2), we may eliminate $E[M]$ in equation (10.6.5), therefore giving

$$E[N_j] = 1 + \lambda(1 - \frac{1}{\rho}) \qquad (10.6.6)$$

2. Dynamic equation for N_j

We can derive another expression for $E[N_j]$ by considering the state dynamics of N_j. The next state N_j' is given by

$$N_j' = N_j - \epsilon(N_j) + A_j \qquad ;\text{where } \epsilon(N_j) = \min(1, N_j) \qquad (10.6.7)$$

In which A_j is the number of fresh HOL packet arrivals at the M unblocked inputs for the output j. Notice that this equation has the same form as that for the M/G/1 queue of equation (10.2.1). Now A_j is relatively independent of N_j provided that N, the total number of inputs, is large. Applying the mean-value equation (10.2.8), we have

$$E[N_j] = E[A_j] + \frac{E[A_j(A_j - 1)]}{2(1 - E[A_j])} \qquad (10.6.8)$$

This expression for $E[N_j]$ is purely a function of the first and second moments of A_j. The first moment $E[A_j]$ is simply λ whereas $E[A_j(A_j-1)] = \lambda^2$. Intuitively, this seems correct since A_j is a Poisson random variable with mean $(E[A_j])$ and variance $(E[A_j^2] - E[A_j]^2)$ both equal to λ by Campbell's theorem. We derive these moments as follows, which may be skipped if the reader is already convinced.

3. Computing first and second moments of A_j

A_j is the number of fresh HOL arrivals at the M unblocked inputs for output j. Due to flow conservation, the mean of A_j must be equal to the total arrivals at the input for j in steady state, hence

$$E[A_j] = \lambda \qquad (10.6.9)$$

Each of these M inputs may have a fresh HOL packet with probability ρ, and the destination address of the packet is equally distributed for each of the N outputs. Hence, each of the M inputs

10.6 HOL Effect for Packet Switches

independently contributes, with probability ρ/N, a HOL arrival for A_j. Thus the binomial distribution of these M 0-1 random variables has a z transform

$$F_M[z] \stackrel{\Delta}{=} E[z^{A_j} | M] = \left[1 - \frac{\rho}{N} + \frac{\rho}{N}z\right]^M \qquad (10.6.10)$$

If we differentiate $F_M[z]$ twice with respect to z, and let $z=1$, we obtain:

$$E[A_j(A_j-1)|M] = \rho^2 \frac{M(M-1)}{N^2} \approx \left(\rho \frac{M}{N}\right)^2 \qquad (10.6.11)$$

for large M. It seems that the above expression depends on M, which in turn depends on N_j. However, by applying the law of large number, M/N should converge to a constant for large N. Using the flow conservation equation $E[M]\rho = N\lambda$, equation (10.6.11) becomes

$$E[A_j(A_j-1)] = \lambda^2 \qquad (10.6.12)$$

for large N.

4. Solution for q and ρ as a function of λ

Using the derived moments, equation (10.6.8) becomes

$$E[N_j] = \lambda + \frac{\lambda^2}{2(1-\lambda)} = \frac{(2-\lambda)\lambda}{2(1-\lambda)} \qquad (10.6.13)$$

Let us now compute q, the probability that a HOL packet is served. Per slot, we have a throughput of $N\lambda$ for the N outputs. The average total number of contending packets for all inputs is $N \times E[N_j]$. Hence

$$q = \frac{N\lambda}{N\,E[N_j]} = \frac{2(1-\lambda)}{2-\lambda} \qquad (10.6.14)$$

Notice that the calculation of q depends entirely on the dynamic equation (10.6.7) for N_j. This fact will be exploited when we study queueing for multi-cast packet switching.

Let us now compute ρ, the packet arrival rate into unblocked HOL positions. Substituting equation (10.6.13) into equation (10.6.6) gives

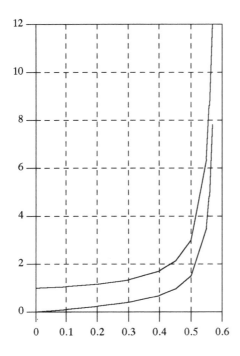

Figure 10. Mean Delay and Queue Length Versus Arrival Rate

$$\rho = \frac{2\lambda(1-\lambda)}{2-2\lambda-\lambda^2} \tag{10.6.15}$$

Equation (10.6.15) relates the input traffic rate λ to the degree of saturation ρ, which measures traffic arrival rate conditioned on a successful removal of the HOL packet in the previous slot.

The maximum λ is obtained at full saturation, namely when $\rho=1$. Solving for λ in equation (10.6.15) when $\rho=1$ gives $\lambda=2-\sqrt{2}=0.586$. Therefore, HOL blocking limits the maximum throughput to 58.6%. Also, there is a noticeable difference between 58.6% and 63.2% obtained by ignoring the HOL correlation of address and occupancy.

10.6 HOL Effect for Packet Switches

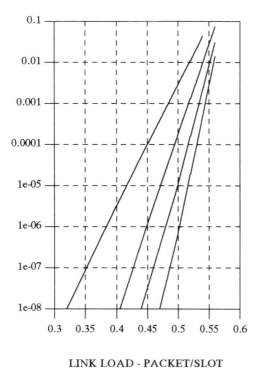

LINK LOAD - PACKET/SLOT

Figure 11. Packet Loss Probability

If packet lengths are assumed to be independently and exponentially distributed, the saturation throughput is 50% with HOL correlation taken into account. This result is left as an exercise.

Given q in terms of λ, we can obtain the mean delay, mean queue length, and the probability of buffer overflow from equations (10.3.13-15), which are plotted in figures 10 and 11.

10.7 Load Imbalance for Single-Stage Packet Switches

In order to obtain small input queueing delay or packet loss probability, the offered load per input has to be reduced from the maximum 58.6%. From figures 10 and 11, delay and loss are reasonably small for an offered load of 50%. In order to handle offered load per input exceeding 50%, we should speed up the switching network relative to the inputs. Therefore, a speed-up factor of 2 is sufficient to handle up to full loading of inputs at 100%. When the switching network is speeded up relative to the outputs, buffering is needed for output queueing. As the switching network is speeded up, the queueing delay shifts from the input queues to the outputs queues. Given sufficiently large speed-up, queueing is shifted altogether to the output queues. For this case, the queueing delay at the output is given by analyzing the output queue, which has a Poisson distribution (rate λ) for the number of packets arriving per slot time, and the deterministic service rate of 1 packet per slot. With uniform and random traffic at both the input and the outputs, the output queueing effect intrinsically forces us to reduce the offered traffic per input (which is also the offered traffic per output) below 100% in order to avoid excessively long output queues.

We analyze here load imbalance, both at the input or at the output. We have shown in this chapter that without a switch speed up, the switch can have a maximum throughput of 58.6%. Suppose the input ports are uniformly loaded at slightly less than 58.6%. In steady state, the input queues have finite expected queue lengths. Suppose we reduce the load at some of the inputs, while still assuming uniform and random output addresses. It is obvious that the input queues cannot suffer from this reduction of load. Let us speed up the switch by a factor of 2 relative to the inputs. All inputs now effectively have less than 50% load, no matter how imbalanced the input loads are. Hence load imbalance can be properly dealt with by speeding up the switching network.

The analysis for output imbalance generated by non-uniform output addresses is more complicated than that for input imbalance. Let us assume that the input load is balanced across the inputs at a Bernoulli rate of λ packets per slot per input. Furthermore, let us assume that packet arrivals at each queue have independent output addresses. However, the output addresses are not uniformly distributed for the outputs j. Let g_j be the output address imbalance factor such that the probability of an output address being j is equal to g_j/N. Obviously, we must have $\sum_j g_j = N$ in order that the output address probabilities

10.7 Load Imbalance for Single-Stage Packet Switches

sum to 1. The uniform address case considered previously has $g_j = 1$ for all j.

Let us go through the traffic analysis for the HOL system of packet switches once again with output imbalance in mind. The flow conservation equation through the HOL system, namely

$$E[M]\rho = N\lambda \tag{10.7.1}$$

remains the same as equation (10.4.2). Let us examine equation (10.6.3) which governs the dynamics of the HOL system. Now the random variables N_j are no longer symmetrical. However, the sum of the throughput of all outputs divided by N remains λ. Therefore, the corresponding equation for equation (10.6.5) is

$$\frac{\lambda}{\rho} = \frac{E[M]}{N} = 1 - \frac{\sum_j E[N_j]}{N} + \lambda \tag{10.7.2}$$

It remains to compute $E[N_j]$ given by the mean-value equation (10.6.8). In steady state, flow conservation implies

$$E[A_j] = g_j \lambda \tag{10.7.3}$$

For given M, the corresponding formula of equation (10.6.10) is

$$E[z^{A_j} | M] = \left[1 - \frac{\rho g_j}{N} + \frac{\rho g_j}{N} z\right]^M \tag{10.7.4}$$

Subsequently, we obtain in similar fashion

$$E[A_j(A_j - 1)] = g_j^2 \lambda^2 \tag{10.7.5}$$

Using equations (10.7.3) and (10.7.5), we obtain an analogous equation to equation (10.6.13).

$$E[N_j] = g_j \lambda + \frac{g_j^2 \lambda^2}{2(1 - g_j \lambda)} \tag{10.7.6}$$

It is important to check if $E[N_j]$ is substantially smaller than the number of input queues N, because we have assumed in the processes of computing $E[N_j]$ that the arrivals A_j are relatively independent of

N_j. Notice that equation (10.7.6) is independent of N. Therefore the independence assumption is good for large N.

Substituting equation (10.7.6) into equation (10.7.2) and using the fact that the g_j sum to N, we obtain

$$1 - \frac{\lambda}{\rho} = \frac{1}{N} \sum_j \frac{g_j^2 \lambda^2}{2(1-g_j \lambda)} \qquad (10.7.7)$$

subject to the normalizing condition

$$\sum_j g_j = N \qquad (10.7.8)$$

The left hand side of equation (10.7.7) is decreasing in λ, whereas the right hand side is increasing in λ. The two sides intersect at a unique solution for λ in [0,1], if a solution exists.

Let us make several observations concerning equation (10.7.7). First, we must have $g_j \lambda < 1$ for all j. The violation of such a condition for any output implies the output is loaded beyond 1, and consequently the overloaded output causes a serious backlog at all input ports. This disastrous blocking phenomenon can be avoided by three means. First, we may speed up the switching network sufficiently to shift this input backlog to the output queue, which will then becomes excessively overloaded. Thus we isolate the switching network from the intrinsic problem of output overload. Second, we may check (by virtual call setup or empirical observation) output load and apply load control. Third, a packet which repeatedly fails the HOL contention probably indicates some sort of failure (output overload, dead devices) at the output requested by the HOL packet. The input should not be tied up by such failures. Packets behind the blocked HOL packet should be able to bypass the HOL packet.

The second observation involves changing the load imbalance factors g_j subject to the normalizing equation (10.7.8). The right hand side of equation (10.7.7) is a concave function in each g_j. In other words, a linear combination of two sets of g_j each satisfying equation (10.7.9) produces a functional value which is larger than the corresponding linear combination of the functional values at the two sets of g_j. Consequently, λ is maximized when $g_j = 1$ for all j. In other words, output imbalance always reduces total throughput of the network.

10.7 Load Imbalance for Single-Stage Packet Switches

As an exercise, consider the case that half of the outputs have twice the load of the remaining half. Hence the load imbalance factors are respectively 4/3 and 2/3. Throughput is maximized at saturation with $\rho=1$. The resulting λ from equation (10.7.7) is 52.9%, which is less than 58.6% for the balance load case by 5.7%. Multiplying λ by the load imbalance factors, the loads for the two types of outputs are 35.3% and 70.6%. Consequently, it is possible to have individual outputs loaded beyond 58.6%.

In conclusion, the output imbalance problem does not severely affect throughput, provided that the output load is less than the saturation throughput calculated for given g_j. It is always prudent to have a switch speed-up of 2 or more as a safe-guard for both input and output imbalance.

10.8 Queueing for Multi-Cast Packet Switches

For multi-cast packet switches, a packet arriving at an input queue can be destined for a random number F of outputs. Let r_f be the probability that there are $F=f$ destinations for a packet. We call f the fanout of a packet. These f destinations are assumed distinct and uniformly distributed for the outputs. The number N of inputs (and outputs) for the packet switch is assumed to be much larger than the maximum number of destinations per packet. Hence we may approximate the destinations to be independently and uniformly distributed for all outputs.

Only the HOL packet for each input is served. In each time slot, some destinations are served according to the availability of the outputs. The remaining destinations of a packet contend for delivery to the outputs in the next time slot. When all destinations of a packet are served, the HOL packet is served and a fresh packet may move into the HOL position. We say the fanout is split in this case. Alternatively, we may not allow fanout splitting by insisting that all destinations have to be served in the same time slot. Obviously, fanout splitting provides better throughput and smaller delay for multi-cast packet switching. We shall assume fanout splitting in the following analysis.

The throughput λ is measured per output. Assuming the packet arrival rate per input per slot to be λ_{input}, we have

$$\lambda = E[F]\lambda_{input} \quad ; \text{ where } E[F] = \sum_{f=0}^{\infty} f \, r_f \qquad (10.8.1)$$

Similar to input queued point-to-point packet switching, we assume that each HOL destination is served independently (from slot to slot and across the inputs) with probability q. Recall that in computing q for the point-to-point case (equations (10.6.7 - 14)), q depends only on the HOL dynamic equation (10.6.7) for N_j. Since the HOL dynamics is not changed for the multi-cast case, we have the same conclusion as equation (10.6.14).

$$q = \frac{2(1-\lambda)}{2-\lambda} \qquad (10.8.2)$$

Therefore, q is independent of the random fanout F. This independence has been confirmed by simulation for large N. (In fact, the independence was first observed via simulation. The simple proof of independence came as a hindsight.)

Knowing q, let us now consider the dynamics of fanout splitting and input queueing. For a fresh HOL packet arrival, the probability that a destination is served in $U=u$ time slots is

$$P_U(u) = q(1-q)^{u-1} \quad ; \quad u \geq 1 \qquad (10.8.3)$$

Consequently,

$$P_U(U>u) = \sum_{k=u+1}^{\infty} q(1-q)^{k-1} = (1-q)^u \qquad (10.8.4)$$

Let U_l be the random number of time slots required before the l-th destination is served, $1 \leq l \leq F$. The packet is served after X time slots, where

$$X = \max_{l=1 \text{ to } F} (U_l) \qquad (10.8.5)$$

Consider now $F=f$. Since the U_l are assumed independent for all $1 \leq l \leq f$, we have

$$P_X(X \leq x \mid F=f) = \prod_{l=1}^{f} P_{U_l}(U_l \leq x) = \left[1-(1-q)^x\right]^f \qquad (10.8.6)$$

Consequently,

$$P_X(X=x \mid F=f) \qquad (10.8.7)$$

10.8 Queueing for Multi-Cast Packet Switches

$$= P_X(X \leq x \mid F=f) - P_X(X \leq x-1 \mid F=f)$$

$$= \left[1-(1-q)^x\right]^f - \left[1-(1-q)^{x-1}\right]^f$$

The probabilities for the service time for a HOL packet are given by the following unconditioning.

$$P_X(X=x) = \sum_f r_f \, P_X(X=x \mid F=f) \tag{10.8.8}$$

The mean service time is given by

$$E[X] = \sum_f r_f \sum_{x=0}^{\infty} x \left(\left[1-(1-q)^x\right]^f - \left[1-(1-q)^{x-1}\right]^f \right) \tag{10.8.9}$$

By performing a binomial series expansion for the powers of the arguments inside the square brackets, and then by exchanging the order of summation for the series with that for x, we obtain

$$E[X] = \sum_f r_f \sum_{k=1}^{f} \binom{f}{k} \frac{(-1)^{k+1}}{1-(1-q)^k} \tag{10.8.10}$$

Using similar algebraic manipulations, the transform formula for X is

$$\mathbf{X}[z] = \sum_f r_f \sum_{k=1}^{f} \binom{f}{k} (-1)^{k+1} \frac{1-(1-q)^k}{(1-q)^k} \frac{1}{1-(1-q)^k z} \tag{10.8.11}$$

Let us now compute the saturation throughput for the case of deterministic fanout, namely that $r_f = 1$ for only one f. At saturation, the input queue length becomes unbounded. According to the Pollaczek-Khinchin mean-value equation (10.2.11), this occurs when

$$1 = \lambda_{input} E[X] \tag{10.8.12}$$

Substituting λ_{input} in equation (10.8.1) and $E[F]=f$ into equation (10.8.3), we have

$$f = \lambda E[X] = \lambda \sum_{k=1}^{f} \binom{f}{k} \frac{(-1)^{k+1}}{1-(1-q)^k} \tag{10.8.13}$$

in which the last equality follows from applying equation (10.8.10).

The value of q in terms of λ is given in equation (10.8.2). Hence, we can solve for λ from equation (10.8.13), giving the following saturation throughputs versus fanouts.

f	1	2	4	8	16
λ	.59	.69	.78	.85	.90

For $f=1,2$, the throughput λ agrees almost exactly with the simulation. For higher values of f, λ agrees with the simulation with at most a 2% difference. Therefore, the assumption of independent service probability q per destination is fairly accurate.

Given q and the distribution for X, we can also use the Pollaczek-Khinchin mean-value and transform equations to compute the mean and the tail distribution for the input queue statistics.

10.9 Exercises

1. Find the mean number of customers in an M/M/1 queue with mean interarrival time $1/\lambda$, mean service time $1/\mu$. Assume that a customer waiting in the queue and is not being served may leave the queue with a probability $\mu \delta t + o(\delta t)$ in any period δt of time.

2. Let us study the mean residue time of a renewal process. Assume that the durations Y between successive renewals are independently distributed with probability density function $g(y)$. Now sample the process in steady state at a random time. Let S be the duration between the sampling instant and the next renewal. S is known as the residual time of the renewal process.

   a. Show that the probability density $f(y)$ of having the sampling instant in a renewal epoch with duration Y is given by

   $$f(y) = \frac{yg(y)}{\int_0^\infty yg(y)dy}$$

b. For given Y, argue that S is uniformly distributed in the interval $[0,Y]$. Hence by unconditioning over Y, obtain

$$E[S] = \frac{\int_0^\infty \frac{1}{2} y^2 g(y)\, dy}{\int_0^\infty y\, g(y)\, dy} = \frac{E[Y^2]}{2E[Y]}$$

c. Obtain $E[S]$ and $E[Y]$ for deterministic $Y=d$, namely when renewals are separated by a fixed duration d.

d. Obtain $E[S]$ and $E[y]$ for exponentially distributed Y with mean d. Explain the counter-intuitive result that the two values are identical.

3. For the M/G/1 queue, use an intuitive argument to show the following:

   a. The probability that server is idle is $1-\lambda E[X]$, where X is the service time for a customer in a queue.

   b. The expected duration for an idle period is $\frac{1}{\lambda}$.

   c. From these two results, show that the expected duration of a busy period is $E[X]/(1-\lambda E[X])$.

4. We derive here a more intuitive proof for the Pollaczek-Khinchin mean-value formula (equation (10.2.11)) for M/G/1 queues.

   a. Suppose there are $Q \geq 0$ be the number of packets waiting in a queue (excluding the packet being served) at the arrival instant of a tagged packet. Let R be the residual time in steady state for completing the work for the packet being served. Show that in steady state, the expected amount of time W the tagged packet spent waiting in the queue before being served is

$$E[W] = E[R] + E[Q]E[X]$$

   where

$$E[X] = \int_0^\infty x b(x)\, dx$$

is the expected service time per packet according to the service duration density function $b(x)$.

b. Apply Little's theorem in equation (10.1.8) to show that

$$E[W] = \frac{E[R]}{1-\lambda E[X]}$$

c. We now focus on computing $E[R]$. As defined in the text, let K be the number of packets in the system, including the packets waiting in the queue and the packet being served. Derive the following steps:

$$E[R] = P(K=0) \times 0 + P(K>0) \times E[R \mid K>0]$$

$$= \lambda E[X] \times \frac{E[X^2]}{2E[X]}$$

$$= \frac{1}{2}\lambda E[X^2]$$

(Hint: Use results from the previous problems.)

d. Combining the previous results, obtain the Pollazcek-Khinchin equation in the following form.

$$E[K] = \lambda E[X] + \frac{\lambda^2 E[X^2]}{2(1-\lambda E[X])}$$

which is equation (10.2.11).

5. Consider the Pollaczek-Khinchin transform equation (10.3.4) with the Bernoulli random variable γ being 1 with probability 1. Hence the Pollaczek-Khinchin transform equation becomes

$$\mathbf{K}[z] = p_0 E[z^\alpha]\frac{1-z^{-1}}{1-z^{-1}E[z^\alpha]}$$

Use this equation to derive the mean and variance of K for

a. The M/M/1 queue.

b. The M/G/1 queue. (The mean should agree with the Pollaczek-Khinchin mean value formula.)

10.9 Exercises

6. Consider the problem of bulk arrival for an M/G/1 queue. At an arrival instant, there is a probability h_l of having l customers arriving simultaneously in a group. Let $\mathbf{H}[z] = \sum h_l z^l$.

   a. Show that the number of customers arriving during an interval of length t has a z-transform given by $e^{-\lambda t(1-\mathbf{H}[z])}$.

   b. Show that the number of customers arriving during the service interval of a customer has a z-transform given by $\psi_B(-\lambda(1-\mathbf{H}[z]))$. (Hint: See equation (10.3.7).)

   c. Show that the moment generating function for the number of customer in the system is
   $$\mathbf{K}[z] = p_0 \frac{(1-z^{-1})\psi_B(-\lambda(1-\mathbf{H}[z]))}{1 - z^{-1}\psi_B(-\lambda(1-\mathbf{H}[z]))}$$

   d. Find an expression for p_0, the probability of finding an empty queue.

7. Let us investigate the saturation throughput of an $N \times N$ slotted packet switch with multiple rounds of arbitration, which was described in section 6.4. We shall ignore the dependence of addresses resulting from the arbitration process. (For one round of arbitration, we have shown in section 10.6 a saturation throughput of 0.632 when the dependence is ignored, versus 0.586 when the dependence is accounted for.)

   a. Let $g_k N$ be the number of outputs assigned an input after k rounds of arbitration. Prove that the probability that an unassigned output becomes assigned at the $k+1$ round of arbitration is
   $$1 - \left(1 - \frac{1}{N}\right)^{(1-g_k)N} \approx 1 - e^{-(1-g_k)}$$

   Hence obtain the recursion
   $$1 - g_{k+1} = (1 - g_k)e^{-(1-g_k)}$$

   b. Show that
   $$g_k = 1 - \exp(\exp(\cdots(\exp(-1))))$$

where the exponentiation (base e) is taken k times. Evaluate g_k for $1 \le k \le 8$.

8. Consider an unslotted packet switch with N inputs and N outputs. Assume that packet lengths are independently and exponentially distributed with mean $1/\mu$. Also, assume that the input queues are saturated. When a HOL packet is served, the next packet moves to the HOL position. It is served by the requested output if the output is free; otherwise it is considered blocked. Upon the completion of service for a HOL packet, an output randomly chooses to serve a blocked HOL packet destined for that output.

   a. Draw the Markov birth-death state diagram for N_j, the number of HOL requests for the output j. (Hint: Assume that a constant fraction f of the output servers are busy. These servers, upon completion of service, may contribute an increment for N_j. A decrement for N_j occurs when the server j completes the service for a packet.)

   b. Show that

   $$E[N_j] = N\frac{f}{1-f}$$

   Show that with saturated input queues, $f=0.5$. (Hint: consider the sum of N_j for all j at saturation.) Hence argue that the saturation throughput is 0.5.

9. Consider the case when there are $l>1$ servers per output for an unslotted packet switch. The $lN \times lN$ packet switch has its outputs grouped into N trunk groups each with l trunks. (A method for realizing a slotted and trunk grouped packet switch was discussed in problem 9 of chapter 6.) We shall evaluate the saturation throughput for the switch. Similar to the previous problem, packet lengths are assumed to be independently and exponentially distributed with mean $1/\mu$.

   a. For $l=2$, draw the Markov chain for N_j, the number of HOL packets for the trunk group j. Let $f \times lN$ be the number of busy servers.

   b. Evaluate the steady state probabilities for the Markov chain, and then show that

10.9 Exercises

$$E[N_j] = \frac{4lf}{(2+lf)(2-lf)}$$

c. Show that the saturation throughput is $f = (\sqrt{5}-1)/2 = .618$.

d. (Optional) We have found $f=.5$ for $l=1$ and $f=.618$ for $l=2$. More generally, show that f at saturation satisfies the following equation:

$$(1-f)\left[1 + \frac{lf}{1!} + \cdots + \frac{(lf)^{l-1}}{(l-1)!}\right] = \frac{(lf)^l}{l!}\left[\frac{f}{(1-f)^2} - 1\right]$$

Solving this equation, we obtain

$l =$	3	4	5	6	7	8	16
$f =$	.678	.715	.742	.762	.778	.792	.849

10. Consider a slotted packet switch with lN inputs and N outputs. Show that the saturation throughput is given by $(1+l) - \sqrt{1+l^2}$. (Hint: Use equation (10.6.13) and also consider in saturation the sum of N_j for all j.)

11. Rank the saturation throughput per output for the following packet switches:

   a. A slotted $lN \times N$ packet switch.

   b. A slotted $N \times N$ packet switch with l times speed-up.

   c. A slotted $lN \times lN$ packet switch with output trunk groups of size l. (The slotted and trunk grouped packet switch has a higher saturation throughput than the unslotted and trunk grouped packet switch considered in the previous problem. For $l=16$, simulation results show a throughput of 0.883, versus 0.849 derived in the previous problem for the unslotted switch.)

   Explain the logic behind your ranking.

10.10 References

Kleinrock, in his two volume text on queueing theory, pioneered the use of this theory for computer communication networks. The description in this chapter, particularly that on the M/G/1 queue is fashioned after his presentation. A more succinct treatment of queueing theory than Kleinrock's can be found in the textbook by Bertsekas *et. al.*

The bulk of this chapter analyzes queueing for multi-input multi-output packet switching systems. The analysis of packet multiplexers can be found in the paper by Beckner *et. al.* The analysis of queueing delay at the input as a function of the offered traffic λ was studied in the paper by Hui *et. al.*, using the state space decomposition method described in this chapter. The paper by Karol *et. al.* also derived the saturation throughput by a different method. The analysis on imbalanced traffic is new, and an independent result can also be found in the paper by Li. The queueing analysis for multicast packet switching is also new, which will be reported in more details in a forthcoming paper.

1. M. W. Beckner, T. T. Lee, and S. E. Minzer, "A protocol and prototype for broadband subscriber access to ISDNs," in *Conference Proceeding, Int. Switch. Symp.*, Phoenix, AZ, March 1987.

2. D. Bertsekas and R. Gallager, *Data Networks, Chapter 3*, NY: Prentice Hall, 1987.

3. J. Y. Hui and E. Arthurs, "A broadband packet switch for integrated transport," *IEEE J. Selected Areas Commun.*, vol-5, pp. 1274-1283, Oct. 1987.

4. J. Y. Hui and T. Renner, "Queueing analysis for multicast packet switching", submitted to *IEEE J. Selected Areas Commun.*

5. M. J. Karol, M. G. Hluchyj and S. P. Morgan, "Input versus output queueing on a space-division packet switch", *IEEE Trans. Commun.*, vol-35, pp. 1347-1356, Dec. 1987.

6. L. Kleinrock, *Queueing Systems, Vol. 1: Theory*, Wiley Interscience, New York, 1975.

10.10 References

7. S. Q. Li and M. J. Lee, "A study of traffic imbalances in a fast packet switch," *Proc. of Info. Com.,* vol. 2, pp. 538-547, April 1989.

CHAPTER 11

Queueing for Multi-Stage Packet Networks

- 11.1 Multi-Stages of M/M/1 Queues
- 11.2 Open and Closed Queueing Networks
- 11.3 Application to Multi-Stage Packet Switching
- 11.4 Analysis of Banyan Networks with Limited Buffering
- 11.5 Local Congestion in Banyan Network
- 11.6 Appendix- Traffic Distribution for Permutations

In chapter 9, we formulated the multi-stage circuit switching problem as a multi-commodity resource allocation problem, for which a terminal pair requests the simultaneous allocation of bandwidth on a set of links. In contrast, packet switching does not require simultaneous allocation but allocation on a link-by-link basis. Local conflicts in resource allocation result in queueing. Based on results for single stage queueing developed in the previous chapter, we approximate the queueing behavior in multi-stage packet networks in this chapter. The first two sections deal with more general results such as reversibility and product form solutions for networks. The last three sections discuss performance analysis methods for buffered banyan networks under different traffic assumptions.

11.1 Multi-Stages of M/M/1 Queues

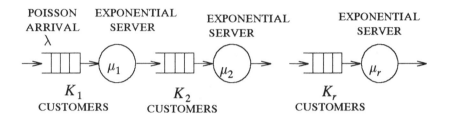

Figure 1. Tandem Queues

In this section, we consider r queues in tandem as shown in figure 1, in which the ith queue has exponential service time distribution with mean $1/\mu_i$. We assume that customers arrive at the first queue with exponential interarrival time distribution of mean $1/\lambda$. Each customer requests an independent amount of service from each queue. We would like to answer two questions:

1. How long will a customer sojourn in the system?
2. What is the number of customers at each queue during a snapshot?

The answers to these two questions are amazingly simple - just treat the problem as though the queues are totally independent, even though that the arrival processes to each queue are not independent. Furthermore, the arrivals to each queue are Poisson, not just for the first queue. Let the random variable T_i be the sojourn time of a customer for the ith queue in tandem. Let the random variable K_i be the number of customers in the ith queue at a snapshot. Then the answers to the above questions are given as follows. The probability density for the times spent at the nodes is

$$p(T_1=t_1,\cdots,T_r=t_r) = \prod_{i=1}^{r} p(T_i=t_i) = \prod_{i=1}^{r} \left(\mu_i-\lambda\right) e^{-(\mu_i-\lambda)t_i} \quad (11.1.1)$$

The probability for the distribution of customers in the nodes is

11.1 Multi-Stages of M/M/1 Queues

$$P(K_1=k_1, \cdots, K_r=k_r) = \prod_{i=1}^{r} P(K_i=k_i) = \prod_{i=1}^{r} \left(1 - \frac{\lambda}{\mu_i}\right)\left(\frac{\lambda}{\mu_i}\right)^{k_i} \quad (11.1.2)$$

These product form solutions suggest certain independency among the queues in tandem. It seems that each queue behaves like an M/M/1 queue independently of the other queues. To show this, we have to prove two facts. First, the departure process of an M/M/1 queue, which forms the arrival process to the next queue in tandem, is Poisson. Second, the state of an M/M/1 queue is independent of its past departure process. Since the state of the next queue in tandem is determined by the departure process of the previous queue up to the moment considered, the states of the two queues in tandem are independent. These results are known as Burke's Theorem.

Before we state and prove Burke's Theorem more formally, we have to introduce the notion of a reversed process and the property of reversibility. We discussed reversibility briefly in chapter 8.

Let the set of states be S, and consider a continuous or discrete time Markov process $S(t)$ in steady state. Consider a state trajectory $s_1, s_2, \cdots, s_n$ for which the transitions to these states occur at times $\tau_1, \tau_2, \cdots, \tau_n$ respectively. The reversed state trajectory is defined by the state sequence $s_n, s_{n-1}, \cdots, s_1$ for which the transitions to these states occur at times $\tau - \tau_n, \tau - \tau_{n-1}, \cdots, \tau - \tau_1$ for a given τ. The forward and reversed trajectories are illustrated in figure 2. We call $S(-t)$ the reversed process for the forward process $S(t)$.

A stochastic process is reversible if the forward trajectory and the reversed trajectory have the same probability for all trajectories and τ. Reversibility implies that in steady state, we cannot distinguish statistically the forward process from the reverse process. A film for such processes would appear normal when played backward.

As an example of a reversible process, consider the M/M/1 queue. Suppose we enclose the M/M/1 queueing system in a black box and make a film recording the arrivals and departures to the black box. Now let us run the film in reverse, and treat the input to the black box as the output, and vice versa. Another way to visualize the reversed process is to use a mirror to obtain the mirror image of the forward process as shown in figure 3. The M/M/1 queue is reversible in the sense that we cannot tell distinguish whether the right hand side or the left hand side is the forward process.

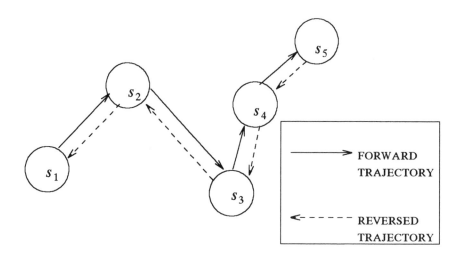

Figure 2. Forward and Reversed Trajectories for Markov Process

Let us examine figure 2 again to see the relationship between reversibility and the concept of detailed balance introduced in chapter 8. Recall that a Markov process has the detailed balance property if for every pair of states $s, s' \epsilon S$, we have

$$p_s p_{ss'} = p_{s'} p_{s's} \qquad (11.1.3)$$

for the steady state probabilities $p_s, p_{s'}$ and the transition probabilities $p_{ss'}, p_{s's}$. This condition guarantees that in steady state, the probability flux exchanged between two adjacent states are balanced. Since the reversed process uses the reversed path for the forward process, it seems then the probability flux along the paths used by both processes should balance each other if detailed balance is satisfied. More rigorously, we are going to show that a Markov process is reversible if and only if it satisfies detailed balance.

Proof:

11.1 Multi-Stages of M/M/1 Queues

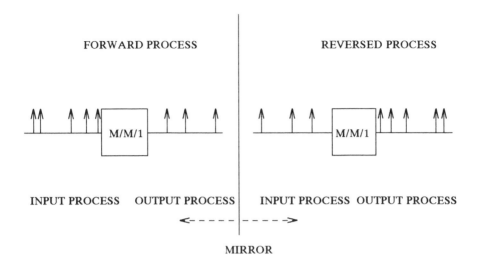

Figure 3. Statistical Symmetry for M/M/1 Queue

Consider the trajectory of states $s_1, s_2, \cdots, s_n$ with transition times $\tau_1, \tau_2, \cdots, \tau_n$ respectively. The probability densities of the forward and reverse processes are products of three factors:

1. The initial probabilities p_{s_1} for the forward process and p_{s_n} for the reverse process.

2. The product of transition probability densities for the associated transitions, namely $\prod_{i=2}^{n} p_{s_{i-1} s_i}$ for the forward process, and $\prod_{i=n}^{2} p_{s_i s_{i-1}}$ for the reverse process.

3. The product of probability densities of remaining in a state between transition times. For both the forward process and the reverse process, these densities are the same since they stay in each state for the same amount of time.

Therefore, we must have

$$p_{s_1}\prod_{i=2}^{n}p_{s_{i-1}s_i} = p_{s_n}\prod_{i=n}^{2}p_{s_is_{i-1}} \qquad (11.1.4)$$

in order to satisfy reversibility. If we apply the detailed balance equation to the right hand side of equation (11.1.4), we have

$$p_{s_n}\prod_{i=n}^{2}p_{s_is_{i-1}} = p_{s_n}\,p_{s_ns_{n-1}}\prod_{i=n-1}^{2}p_{s_is_{i-1}} \qquad (11.1.5a)$$

$$= p_{s_{n-1}}p_{s_{n-1}s_n}\prod_{i=n-1}^{2}p_{s_is_{i-1}} \qquad (11.1.5b)$$

$$= p_{s_1}\prod_{i=n}^{2}p_{s_{i-1}s_i} \qquad (11.1.5c)$$

with equation (11.1.5b) resulting from the detailed balance equation (11.1.3), and equation (11.1.5.c) resulting from the recursive application of the steps in (11.1.5a,b). Equation (11.1.5c) is simply the right hand side of equation (11.1.4). Hence detailed balance implies reversibility.

Conversely, a reversible process has

$$P\Big(s(t){=}s_1, s(t{+}\delta){=}s_2\Big) = P\Big(s(t){=}s_2, s(t{+}\delta){=}s_1\Big) \qquad (11.1.6)$$

Hence

$$p_{s_1}\frac{P\Big(s(t{+}\delta)=s_2\,|\,P(s(t){=}s_1)\Big)}{\delta} = p_{s_2}\frac{P\Big(s(t{+}\delta)=s_1\,|\,P(s(t){=}s_2)\Big)}{\delta}$$

$$(11.6.7)$$

Letting $\delta\to 0$, we obtain the detailed balance equation $p_{s_1}p_{s_1s_2}=p_{s_2}p_{s_2s_1}$.

A similar reasoning can be used for the discrete time Markov process. □

We have shown in chapter 8 that the multi-dimensional birth-death process with truncation of states satisfies the detailed balance equation. Hence such processes are reversible, including the simple M/M/1 queue for which the states represent the number of customers

11.1 Multi-Stages of M/M/1 Queues

in the system.

Now for the M/M/1 queue, a decrement of the state corresponds to the departure of a customer for the forward process, which also corresponds to the arrival of a customer for the reverse process. Since the statistics of the forward process and the reverse process are indistinguishable in steady state, we have the following conclusions drawn from the equivalent relations ↔ in brackets.

Burke's Theorem

1. The departure process is Poisson. (↔ The arrival process is Poisson for the reverse process.)

2. The number of customers in the queue at time t is independent of the departure process before time t. (↔ The number of customers in the queue at time t is independent of the arrival process after time t for the reverse process.)

3. For a particular customer, the system time T (the length of time between arrival to and departure from the system) is independent of the departure process before his departure. (↔ For a particular customer, the system time T is independent of the arrival process after his arrival for the reverse process. This is due to the simple reason that with FCFS discipline, his system time cannot be affected by later arrivals).

In fact, the answers to the two questions posed at the beginning of this section for r M/M/1 queues in tandem are the direct consequence of these conclusions.

1. How long will a customer sojourn in the system?

    Each queue has memoryless arrivals since the departures from each queue are memoryless by conclusion 1. Hence all queues are M/M/1 queues. For a particular customer, his waiting time in queue 1 is independent of the departures prior to his departure by conclusion 3, hence independent of the arrival process to queue 2 before his departure from queue 1. Therefore, his waiting time in queue 1 is independent of his waiting time in queue 2, which in turn is independent of his waiting time at queue 3, etc. Hence we obtain the product form of equation (11.1.1). The derivation for the density $p_i(T_i)$ is left as an exercise for the readers.

Consider the case of $\mu_i = \mu$ for $1 \leq i \leq r$. The total sojourning time $t_s = t_1 + \cdots + t_r$ in the r queues can be derived using equation (11.1.1) by integration. We leave as an exercise for the reader to show that

$$p(T=t_s) = \frac{(\mu-\lambda)((\mu-\lambda)t_s)^{r-1} e^{-(\mu-\lambda)t_s}}{(r-1)!} \qquad (11.1.8)$$

2. What is the number of customers in the system?

The number of customers at time t in queue 1 is independent of the departure process before t by conclusion 2. The departure process determines the number of customers at time t at queue 2. Applying this argument to all queues, we obtain the product form of equation (11.1.2).

Burke's theorem can be generalized for only a few cases. The result remains valid for M/M/m queues, instead of M/M/1 queues. For general inter-arrival or service time distributions, Burke's theorem is usually not true. However, it is true for the particularly important case of *Geom/Geom/1* queues in tandem. In a communication network context, a *Geom/Geom/1* queue corresponds to queues of fixed length packets, for which the arrival and service probabilities per slot are both Bernoulli. We assume these probabilities to be p and q respectively.

Proof:

We may argue that the discrete time Markov chain shown in figure 4 for a *Geom/Geom/1* queue satisfies the detailed balance equation, hence the state trajectory is reversible. However, the departure process cannot be uniquely specified given the state trajectory, since remaining in a state can result from either no packet arrival and departure, or a packet arrival and departure. To resolve this ambiguity, we introduce the inconsequential artifact of staggering the departure times half a slot time behind the arrival times. The corresponding discrete time Markov chain is shown in figure 5.

The resulting discrete time Markov chain satisfies detailed balance, as all birth-death process always do. Therefore, the process is reversible and we cannot distinguish statistically the arrival and the departure processes. Hence Burke's theorem remains valid. □

11.1 Multi-Stages of M/M/1 Queues

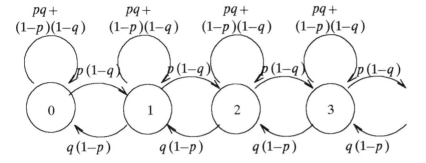

Figure 4. The Discrete Time Markov Chain for Service of Fixed Length Packets

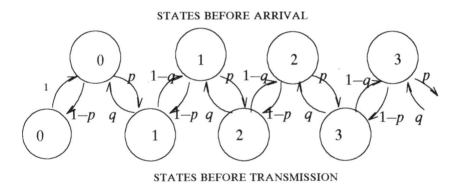

Figure 5. Markov Chain for Arrivals and Departures of Fixed Length Packets

The preservation of Bernoulli statistics from queue to queue for fixed length packets is rather pleasing for designing ATM networks, at least for the above idealized tandem queue case. Provided the packet arrivals are Bernoulli to start with, we do not have to worry about snowballing effects of generating large clusters of packets through the network. Such clusters may cause serious delay problems if left unchecked.

11.2 Open and Closed Queueing Networks

Having dealt with the rather simple tandem of M/M/1 queues, we consider more general networks. These networks may have a multiplicity of paths connecting any two end points. We denote the graph by (V,E), in which V is the set of nodes, and E is the set of directed edges (v,v') linking pairs of nodes $v,v' \epsilon V$. We assume that the service time for customers at each node is independently and exponentially distributed.

We distinguish between cyclic networks and acyclic networks. An acyclic network is one for which no path in the network visits any node more than once. Otherwise, the graph is considered cyclic.

Having described the structure of the network, we also have to characterize the flow of customers in the network. Here, we shall adopt Jackson's model. Jackson's model assumes that customers enter each node $v \epsilon V$ at a Poisson rate λ_v. The customers are served at each node with an exponential distribution of service time of mean $1/\mu_v$. The amount of service requested is independent from node to node. Upon completion of service from node v, the customer moves to node v', $(v,v') \epsilon E$, with probabilities $f_{vv'}$. The customer may leave the network for good with probability g_v. For these probabilities, we have

$$g_v + \sum_{v':(v,v')\epsilon E} f_{vv'} = 1 \qquad (11.2.1)$$

For an acyclic graph, let us look at each path in the network as a tandem queue, which was studied in the previous section. Each queue in the tandem is fed not only by the output of the previous queue in the tandem, but also customers from other tandem queues which happen to share the same queueing nodes. Similarly, a customer departing from a queue may not go to the next queue in the tandem considered.

The acyclic property of the graph does not allow a customer to go back to earlier queues in the tandem queue he is traversing. Hence Burke's theorem applies and we may use the product form solution for the waiting time distributions and the number of customers in the

11.2 Open and Closed Queueing Networks

system. What remains to be determined is the total traffic rate α_v entering the node v. By considering the conservation of flow, α_v can be derived in terms of the entry rate λ_v and the branching probabilities $f_{v'v}$.

$$\alpha_v = \lambda_v + \sum_{v':(v',v)\in E} f_{v'v}\alpha_{v'} \quad \text{for all } v \in V \quad (11.2.2)$$

Using these linear equations, we can solve uniquely for α_v. For given α_v, the queueing behavior at each node is uniquely determined.

Let us now consider cyclic networks. There are two kinds of cyclic networks. The first kind is open, for which customers arrive at and depart from the network at certain nodes. In other words, there exist nodes v such that $g_v > 0$ (see equation (11.2.1)) and nodes v' such that $\lambda_{v'} > 0$ (see equation (11.2.2)). If the arrivals to the network at each v are Poisson, then the departure processes for v' are independent and Poisson. An outline of the proof for this result is given in the exercises. The Markov property of the reverse process is employed in the proof.

The second kind of cyclic network is closed, for which no arrival to or departure from the network occurs. There is a constant number of customers in the network. For closed cyclic networks, the arrival process at each node is not necessarily Poisson. The non-Poisson character of the circulating component is evident by considering a closed network with $\lambda_v = 0$ for all $v \in V$ and only 1 circulating customer in the whole network. Obviously, interarrival times are not independent, as shown in the exercises.

For both open and closed cyclic networks, Burke's theorem is no longer valid because the circulating component makes the arrival and departure processes mutually dependent. Surprisingly, the distribution of customers in a cyclic network also has a product form. The number of customers (question 2 posed at the beginning of section 11.1) at each node has the same distribution as that found using the method derived above for acyclic networks. In other words, we solve for α_v using the system of equations (11.2.2), which we may use to compute the distribution of the random number K_v of customers in node v as if queueing at v is independent of the other nodes. The distribution is analogous to equation (11.1.2).

Product Form Solution for Cyclic Networks

Define the random vector $\underline{K}=(K_v)_{v \in V}$. The steady state probabilities of $\underline{K}$ are given by

$$P(\underline{K}=\underline{k}) = \kappa \prod_{v \in V} \left(1-\frac{\alpha_v}{\mu_v}\right)\left(\frac{\alpha_v}{\mu_v}\right)^{k_v} \qquad (11.2.3)$$

where α_v is the solution to equation (11.2.2). For open cyclic network, the k_v can be any non-negative integers and the normalization constant κ is equal to 1. For closed cyclic network, $\sum_v k_v = N$, the number of customers in the network, and κ normalizes the probabilities for the constrained state space.

Proof:

We shall show that the product form solution satisfies the equilibrium equation given by

$$\sum_{\underline{k}'} p_{\underline{k}'} p_{\underline{k}'\underline{k}} = p_{\underline{k}} \sum_{\underline{k}'} p_{\underline{k}\,\underline{k}'} \qquad \text{for all } \underline{k} \qquad (11.2.4)$$

The left hand side of equation (11.2.4) gives the probability flow into $\underline{k}$ from all previous states $\underline{k}'$. The right hand side gives the probability flow from $\underline{k}$ to all next states $\underline{k}'$. Consider the transition $\underline{k}' \to \underline{k}$. There are three types of $\underline{k}'$, resulting from:

1. A new customer enters the network at the node v. Hence k_v increases by 1. According to the product form solution given by equation (11.2.3), we have

$$p_{\underline{k}'} = p_{\underline{k}} \frac{1}{\frac{\alpha_v}{\mu_v}} \qquad (11.2.5)$$

The transition probability is $p_{\underline{k}'\underline{k}} = \lambda_v$, the arrival rate of new customers.

2. A customer departs the network from the node v. Hence k_v decreases by 1. According to the product form solution, we have

$$p_{\underline{k}'} = p_{\underline{k}} \frac{\alpha_v}{\mu_v} \qquad (11.2.6)$$

11.2 Open and Closed Queueing Networks

The transition probability is $p_{\underline{k}'\underline{k}} = \mu_v g_v$.

3. A customer moves from node v' to v. Hence k_v increases by 1 and $k_{v'}$ decreases by 1. According to the product form solution, we have

$$p_{\underline{k}'} = p_{\underline{k}} \frac{\dfrac{\alpha_{v'}}{\mu_{v'}}}{\dfrac{\alpha_v}{\mu_v}} \qquad (11.2.7)$$

The transition probability is $p_{\underline{k}'\underline{k}} = \mu_{v'} f_{v'v}$.

We now substitute these values of $p_{\underline{k}'}$ and $p_{\underline{k}'\underline{k}}$ into the left hand side of equation (11.2.4). The values of $p_{\underline{k}\underline{k}'}$ on the right hand side of equation (11.2.4) can also be evaluated according to the above cases. These substitutions give

$$\sum_{\underline{k}'} p_{\underline{k}'} p_{\underline{k}'\underline{k}} \qquad (11.2.8a)$$

$$= \sum_v \left[p_{\underline{k}} \frac{1}{\dfrac{\alpha_v}{\mu_v}} \right] \lambda_v + \sum_v \left[p_{\underline{k}} \frac{\alpha_v}{\mu_v} \right] \mu_v g_v + \sum_{v,v'} \left[p_{\underline{k}} \frac{\dfrac{\alpha_{v'}}{\mu_{v'}}}{\dfrac{\alpha_v}{\mu_v}} \right] \mu_{v'} f_{v'v} \qquad (11.2.8b)$$

$$= p_{\underline{k}} \sum_{\underline{k}'} p_{\underline{k}\,\underline{k}'} \qquad (11.2.8c)$$

$$= p_{\underline{k}} \left(\sum_v \mu_v g_v + \sum_v \lambda_v + \sum_{v,v'} \mu_v f_{vv'} \right) \qquad (11.2.8d)$$

We now proceed to prove the equality of (11.2.8b) and (11.2.8d).

The common factor $p_{\underline{k}}$ in equations (11.2.8b) and (11.2.8d) can be eliminated. Next, break the equality into two parts. First, we equate the second summation in (11.2.8b) and the second summation in (11.2.8d).

$$\sum_v \alpha_v g_v = \sum_v \lambda_v \qquad (11.2.9)$$

The left hand side is the rate of exit from the network, whereas the right hand side is the rate of entry. In steady state, these two rates must be equal.

For the remaining terms, we equate all terms associated with a particular v in equations (11.2.8b) and (11.2.8d).

$$\frac{1}{\frac{\alpha_v}{\mu_v}} \lambda_v + \sum_{v'} \frac{\frac{\alpha_{v'}}{\mu_{v'}}}{\frac{\alpha_v}{\mu_v}} \mu_{v'} f_{v'v} = \mu_v g_v + \sum_{v'} \mu_v f_{vv'} \qquad (11.2.10)$$

The right hand side reduces to μ_v due to equation (11.2.1). Multiplying both sides by α_v/μ_v, we obtain

$$\lambda_v + \sum_{v'} \alpha_{v'} f_{v'v} = \alpha_v \qquad \text{for all } v \qquad (11.2.11)$$

which is simply the flow conservation equation (11.2.2).

We have shown that the product form solution (11.2.3) satisfies the equilibrium equation. □

Consider the special case of a closed network, namely that there is a fixed number N of customers in the network and that no customer ever enters or leaves the network (in other words, $g_v=0$ and $\lambda_v=0$). The flow conservation equation (11.2.11) becomes

$$\alpha_v = \sum_{v'} f_{v'v} \alpha_{v'} \qquad \text{for all } v \in V \qquad (11.2.12)$$

We may normalize the α_v such that $\sum_v \alpha_v = 1$. Consequently, α_v can be interpreted as the probability of finding a customer in node v, given that there is only one customer in the entire network.

11.3 Application to Multi-Stage Packet Switching

The previous sections are cast in the context of service provided to migrating customers. Product form solutions were obtained by assuming that fresh arrivals to the network have Poisson statistics, and that service distributions at each node are independently and

11.3 Application to Multi-Stage Packet Switching

exponentially distributed. Furthermore, the Jackson network considered assumed random routing for packets. In this section, we first recast the problem for multi-stage packet switching. We then examine the adequacy of the assumptions of the previous sections in the new context.

The cyclic and acyclic networks considered in the previous section assumes a common waiting room for the customers. For packet switching, this may or may not be the case. Many computer communication networks do employ a single processor to buffer and route packets. Hence, we have a single buffer shared by the incoming links to a node, as shown in the top figure of figure 6. On the other hand, we may have a single buffer for each input. An interconnection network then connects these input buffers to the outputs. These implementations were described as fast packet switching in chapter 5. For these implementations, we have input queueing and HOL blocking, which were analyzed in the previous chapter for a single stage. Analyzing such multi-stage networks is very difficult in general. We shall describe in section 11.4 the analysis of one such network, namely the slotted buffered banyan network with a single packet buffer per input to an internal node.

We proposed in chapters 5 and 6 that the interconnection network should be speeded up to reduce the HOL blocking. The input queueing effect is then shifted to the output queues. Assuming that the interconnection network inside each node is substantially faster than the input and output links, each node can be represented by the Jackson network shown in the bottom figure of figure 6. Since the interconnection network is a very fast server, there is no queueing at the input queues. Therefore, we may as well model the input queues as a single shared input queue. After a packet is served by the interconnection network, it branches out to the output queues, each with an output server. Hence a network of such nodes constitutes a Jackson network, and can be analyzed as such.

Having modeled the structure of the network for packet switching, we now examine the assumptions for packet routing, arrivals, and services.

1. The random routing assumption

Random routing assumes that at each node v, packets route independently to the next node v' with prescribed branching probabilities $f_{vv'}$. This assumption is quite good for datagram

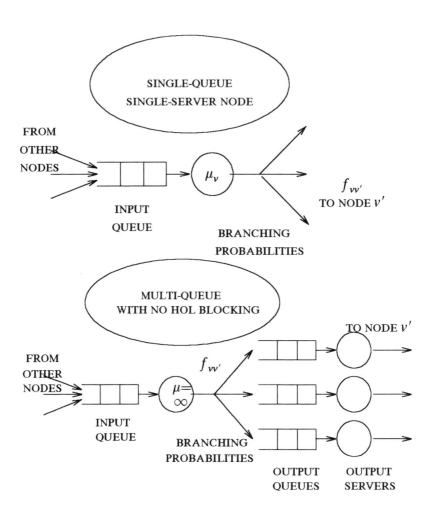

Figure 6. Modeling of Multi-Queue Multi-Server Nodes

networks for which packets are routed on a packet by packet basis. For multiplexed calls with packetized transmission, this assumption is good if each link in the network multiplexes a large number of calls. Otherwise, packet addresses are highly correlated, making analysis difficult.

11.3 Application to Multi-Stage Packet Switching

2. The independent service time assumption

The previous sections assume that each customer requires an independent amount of service from each node. In the packet network context, this assumption is known as Kleinrock's independence assumption, which has been profitably and extensively used for communication network analysis. However, this independence assumption may seem unrealistic since a packet is not expected to change its length from node to node.

Let us consider how this assumption affects the the total sojourning time of a packet in the network. The total sojourning time in the network consists of two components: the queueing time before the packet is transmitted, and the transmission time. The transmission service required, namely the transmission time times the transmission speed, should be the same from node to node. Therefore, the total transmission time for the packet in the network is predictable. On the other hand, queueing time depends on the service times required by other packets in the queue. Since packets follow different routes, the service times of packets in a queue are made random from queue to queue. Hence assuming independent queueing time is reasonable.

3. The exponential service time distribution assumption

The previous sections assume that packet length is exponentially distributed in order to obtain the product form solution for the waiting time distribution and customer distribution in the network. This distribution of packet length can be objectionable, such as for networks transporting fixed length packets. Nevertheless, the exponential assumption is often used because it gives good approximation results and provides insights for network design.

We may broaden the exponential assumption by allowing general packet length distributions. Furthermore, we assume that packet arrivals at each node are Poisson. This Poisson assumption is good if each node has large fan-in and fan-out. In this case, we may treat each node as an M/G/1 queueing system. We first compute the offered traffic α_v to each node according to the flow conservation equation (11.2.2). Next, we compute the queueing behavior per node using formulas for the M/G/1 queue. We then assume that the nodes are independent. This independence is accurate if the fan-in and fan-out of each node is large, and hence a node may not unduly influence another node. Subsequently, we can assume that the distributions of delay from node to node are independent. The total time a packet spends in the network is then given by adding the fixed transmission times to the random and assumed independent queueing times for the nodes enroute.

Having examined these three assumptions in the context of packet switching, we now look at the important special case of fixed length packets for the rest of this section. As shown in section 11.1, the *Geom/Geom/*1 queue, like the M/M/1 queue, is time reversible. Hence Burke's theorem is true for tandem queues serving fixed length packets, so long that the arrival process at the first queue is Bernoulli.

Beyond the tandem queues, we may have multiple (say m) inputs to a queue for general networks. Suppose the arrival process on each input is Bernoulli with an independent arrival with probability p in each slot. Unfortunately, the superposition of these m processes is no longer Bernoulli but binomial. Hence the product form solution for Jackson network does not carry over for the case of fixed length packets, because the Bernoulli property is not preserved from node to node except for tandem queues.

It is rather difficult to analyze a queue with a binomially distributed arrival process. We may approximate the arrival process by either one of the following models. First, we may assume that the number of arrivals per slot is Bernoulli distributed with mean $\lambda = mp$. Alternatively, we may assume that the number of arrivals per slot is Poisson distributed with mean λ. We shall compare the delay for these two models. We shall assume in both case that the HOL packet is served with probability q.

For the Bernoulli model, the total time spent in a node is given by equation (10.3.14)

$$E[T] = \frac{1-\lambda}{q-\lambda} \qquad (11.3.1)$$

For the alternative slotted Poisson arrival model, the z-transform for the number of packets K in the input queue is given by equation (10.3.4), in which γ is the Bernoulli random variable with $E[\gamma]=q$ and α is the Poisson distributed number of arrivals per slot. The z transform for α is given by equation (10.3.7):

$$E[z^\alpha] = e^{-\lambda(1-z)} \qquad (11.3.2)$$

The transform $E[z^{-\gamma}]$ is given by equation (10.3.5). Subsequent

11.3 Application to Multi-Stage Packet Switching

substitutions into equation (10.3.4) give us the transform for K. The mean number of packets in the input queue is given by $E[K]=\mathbf{K}'[z=1]$. We skip the algebra involved in differentiation (preferably performed by computer because of the necessity to use L'Hospital's rule when we set $\mathbf{K}'[z=1]$). Applying Little's result to the end product, we obtain the mean time in a node

$$E[T] = \frac{1-\frac{\lambda}{2}}{q-\lambda} \qquad (11.3.3)$$

Equations (11.3.1) and (11.3.3) give approximately the same values when λ is small, namely in the light traffic regime.

In practice, we choose either one of these models based on the degree of fan-in and fan-out of each node. If the fan-in and fan-out are small, we choose the Bernoulli model; otherwise, we choose the Poisson model. The actual delay performance with binomially distributed arrivals is optimistically estimated by the Bernoulli model, and pessimistically estimated by the Poisson model.

We argued in chapter 6 that it is preferable to have large internally non-blocking nodes for constructing multi-stage packet switched networks, so that the number of stages can be kept small for a network with a given number of inputs and outputs. Otherwise, the large number of stages introduces more queueing delay, which is undesirable. Given the nodes have large fan-in and fan-out, the Poisson model becomes quite accurate. Furthermore, large fan-in and fan-out for the nodes reduce the dependency of queueing among these nodes, and consequently the product form solution for networks of M/G/1 queues discussed in this section can be very accurate.

For transport of fixed length packets, we have analyzed the cases of tandem queues as well as nodes with large fan-in and fan-out. Most other situations are usually difficult to analyze. As an example, we shall analyze the banyan network with 2×2 nodes and limited buffering at each node.

11.4 Analysis of Banyan Network with Limited Buffering

We now proceed to analyze the throughput and delay for routing fixed length packets in a buffered banyan network, which was introduced in chapter 5. The banyan network has 2×2 input buffered nodes, n stages and therefore 2^n inputs and outputs. Each input to the banyan network has an input queue with infinite capacity. However, each internal 2×2 node has limited buffering at its input. Therefore, a HOL packet not only suffers from HOL blocking, but also blocking due to a full buffer at the next stage. We shall assume that each internal input buffer can hold at most one packet.

The state of the network is given by the occupancy of all buffers in the network. Theoretically, we can solve the steady state probabilities using the equilibrium equation for the discrete time Markov chain. However, the large number of states renders this computation infeasible except for very small networks. Instead, we shall introduce the following assumptions for approximations:

1. *Independent buffer occupancy:*

    Each input buffer in the ith stage has a packet independently (of other buffers) with probability r_i.

2. *Independent link occupancy:*

    Conditioned on having a packet in the buffer in the ith stage, the packet can move to the next stage with probability s_i independently (from slot to slot) in the current time slot.

3. *Uniform traffic:*

    The packet moves to one of the two next nodes with equal probability of $\frac{1}{2}$.

Suppose packets arrive at each input of the banyan network at rate λ. Now the product $r_i s_i$ is the rate of flow of packets through an input of an ith stage node. Hence by conservation of packet flow in each stage, we have

$$r_i s_i = \lambda \qquad \text{for all } 1 \leq i \leq n \tag{11.4.1}$$

We can obtain another equation relating traffic flow between successive stages. Consider the probability r_i that a packet is present in the buffer in the current time slot. Either the same packet is blocked in the previous time slot, which occurs with probability $r_i(1-s_i)$; or we have a fresh packet arrival from the $(i-1)$th stage. A fresh arrival may occur only if the buffer is empty at the beginning of the previous slot, which occurs with probability $r_i s_i + (1-r_i)$. (These two terms correspond to the cases that the packet is cleared or the

11.4 Analysis of Banyan Network with Limited Buffering

buffer is not occupied in the first place.) Given an empty buffer, a packet from the $(i-1)$th stage will arrive with probability $r_{i-1}(1-r_{i-1})+\frac{3}{4}r_{i-1}^2$. (These two terms correspond to two cases: either one of the two input buffers at the $(i-1)$th stage is occupied, with probability $1/2$ that its packet moves into a buffer at the ith stage; or both input buffers at the $i-1$th stage are occupied, with probability $3/4$ that one of the two packets ends up in a buffer at the ith stage.) Summarizing the results of these conditionings, we have

$$r_i = r_i(1-s_i) + \left[r_i s_i + (1-r_i)\right]\left[r_{i-1}(1-r_{i-1})+\frac{3}{4}r_{i-1}^2\right] \quad (11.4.2)$$

which after simplification becomes

$$r_i s_i = \left[r_i s_i + (1-r_i)\right]r_{i-1}(1-\frac{1}{4}r_{i-1}) \quad (11.4.3)$$

Substituting $\lambda = r_i s_i$ in equation (11.4.1) into (11.4.3), we may express r_i in terms of r_{i-1} and λ:

$$r_i = 1+\lambda-\frac{\lambda}{r_{i-1}(1-\frac{1}{4}r_{i-1})} \quad (11.4.4)$$

Thus we have a recursive formula for computing the probability buffer occupancy for consecutive stages.

At the last stage ($i=n$), we only have HOL blocking but no blocking due to full buffer because packets arriving at the output of the banyan network are always removed immediately. A HOL packet is blocked with probability $1/2$ if there is a HOL packet for the same output at the other input to the node. The probability that the other HOL packet want the same output is $1/2$. Therefore

$$s_n = 1-\frac{1}{4}r_n \quad (11.4.5)$$

Let us guess a value for given λ. We can solve for r_n in terms of λ by substituting the value of $r_n s_n = \lambda$ into (11.4.5). Solving the resulting quadratic equation, we obtain

$$r_n = 2(1-\sqrt{1-\lambda}) \quad (11.4.6)$$

Knowing r_n, we can solve for r_{n-1} in equation (11.4.4). This process is then applied recursively for smaller values of i. In the process, we have to check that the resulting r_i and s_i are legitimate probabilities in the range [0,1]. Otherwise, the λ assumed is not attainable.

For given n, the maximum λ is obtained when the resulting $r_1=1$, namely that the input queues to the banyan network are saturated. Using the above procedure for computing r_1, we can compute the value of λ versus the degree of saturation r_1 for various n. In the light load region of $\lambda<.3$, the relationship is almost linear. In the heavy load region, the throughput saturates at values which decrease as n increases. The saturation throughput itself seems to converge to a limit of .4 for very large n.

How accurate is this method compared to simulation results? By saturating the inputs to the banyan network, the resulting λ obtained by simulation is typically around 10% smaller than the value obtained by analysis. The analysis is optimistic because the independent service (from slot to slot) assumption ignores the fact that when a packet is blocked, it is more likely to be blocked the next slot time. Nevertheless, we consider the approximation good for its simplicity, and more significantly, its accuracy in the light load region.

We can then use the computed service probabilities s_i to estimate the delay in the entire network. Each input queue to the banyan network with infinite buffer size can be modeled as a *Geom/Geom/*1 queue with arrival rate λ and service rate s_1. A packet at the ith stage node suffers a geometrically distributed (mean $1/s_i$) random delay at that node. The resulting total delay estimate is quite satisfactory for engineering purposes in the light load region.

11.5 Local Congestion in Banyan Network

In applying the banyan network for telecommunication switching or parallel computing purposes, the assumption that the destination addresses are uniformly and independently distributed for the outputs is often not valid.

First of all, the traffic can be unbalanced for the outputs. In parallel computing, the favored outputs are called hot spots. The throughput and delay behavior for this unbalanced traffic scenario can be significantly different from the balanced traffic case, even if the total traffic offered to the network is kept constant. Intuitively, this

11.5 Local Congestion in Banyan Network

deterioration results from the spreading of congestion within the network in the neighborhood of the hot spot. (For the sort-banyan network which is internally non-blocking, internal congestion is eliminated. However, the congestion shows up as HOL blocking.) The region of congestion may affect the routing of packets to other outputs. Relatively little is known analytically about how hot spots affect throughput and delay.

In telecommunication switching, the problem is complicated by the possibility that an arriving call may demand a significant fraction of the link bandwidth. Therefore, the packet addresses, instead of being independent from packet to packet, are highly correlated. Even though we may have sufficient bandwidth to accommodate all calls at the output of the network, it is quite likely that another high bandwidth call using overlapping links may cause severe congestion for the overlapping links. This congestion is not necessarily localized near the outputs of the network, but may occur anywhere in the path used by a call.

As pointed out in chapter 5, these high bandwidth calls make the buffered banyan network looks like a circuit switched network, because bandwidth is more or less dedicated to the calls. Since banyan networks can realize only a very small subset of connection patterns for point-to-point circuit switching, buffered banyan networks are very blocking for high bandwidth calls. To alleviate this problem, the switching network has to be speeded up many folds. We shall give a rough estimate of this speed-up.

To simplify analysis, we consider the offered load to each link for a given connection pattern. We assume that all connections require unit bandwidth and compute the probability distribution of the offered traffic on each link. The probability that the offered load exceeds the capacity of the internal link gives us some idea of the probability of blocking on that link.

Consider the case of one call per input. Let us assume that each call is uniformly destined for the outputs. For this case, we may have more than one call destined for an output. For the output link of a k th stage node in an n stages network, the probability that there are i calls using the link is given by the binomial distribution

$$p_{i,k} = \binom{2^k}{i} \left(\frac{1}{2^k}\right)^i \left(1 - \frac{1}{2^k}\right)^{2^k - i} \tag{11.5.1}$$

Obviously, congestion becomes more severe for links close to the output of the banyan network since more inputs can reach these links. Equation (11.5.1) can be simplified for $2^k \gg i$ and large 2^n:

$$p_{i,k} \approx \frac{1}{i!} e^{-1} \tag{11.5.2}$$

which is practically independent of k near the outputs of the banyan network. Hence the probability tail that the offered traffic exceeds s is approximated by

$$1 - e^{-1} \sum_{i=1}^{s} \frac{1}{i!} \tag{11.5.3}$$

To make this probability tail smaller than 10^{-4}, the value of s (which provides an estimate of the speed-up requirement) must be at least 6.

We may also evaluate $p_{i,k}$ conditioned on each call being connected to a distinct output. In other words, the input-output connection pattern is a permutation. The derivation is given in the Appendix. As shown in chapter 5, congestion is most severe in the stages midway in the network for permutation connections. Using results from the Appendix, the speed-up required for permutation connections is roughly the same as that without the permutation restriction.

11.6 Appendix- Traffic Distribution for Permutations

In chapter 5, the links in a banyan network were represented as follows. Suppose that a packet from input a is destined for output b. Let $(a_1 a_2 \cdots a_n)$ be the binary representation of a and $(b_1 b_2 \cdots b_n)$ be the representation of b. The link used by the packet from stage k to stage $k+1$ can be labeled by the following method, which was described in chapter 5. We concatenate $(a_1 a_2 \cdots a_n)$ and $(b_1 b_2 \cdots b_n)$ and move a window of width n after the k-th bit $(a_1 \cdots a_k [a_{k+1} \cdots a_n, b_1 \cdots b_k] b_{k+1} \cdots b_n)$. The binary number in the window would then be the link label.

Suppose the input-output connection pattern is a permutation. Two input-output connections share the same internal link at the kth stage if they have the same address inside the window defined in the previous paragraph. There are 2^k inputs to the network which can be

11.6 Appendix- Traffic Distribution for Permutations

connected to the internal link considered. Let us call this set $I' \subset I$, in which I is the set of all inputs. There are 2^{n-k} outputs of the network which can be connected to the same internal link. Let us call this set $O' \subset O$, in which O is the set of all outputs.

Let us now count the number of permutations with i input-output connections sharing the internal link considered. First, the number of ways to choose i inputs from I' is given by:

$$F_1 = \binom{2^k}{i} \tag{11.6.1}$$

The number of ways to connect these i inputs to O' is given by:

$$F_2 = \frac{2^{n-k}!}{(2^{n-k}-i)!} \tag{11.6.2}$$

The remainder of I' (with $2^k - i$ elements) has to be connected to $O - O'$ (with $2^n - 2^{n-k}$ elements), since any input in I' connected to any output in O' must use the internal link considered, resulting in more than i inputs connected to that link. The number of ways for these connections is given by:

$$F_3 = \frac{(2^n - 2^{n-k})!}{(2^n - 2^{n-k} - (2^k - i))!} \tag{11.6.3}$$

The number of ways to connect the remaining elements of I to the remaining elements of O (resulting in connections which cannot use the internal link considered) is given by:

$$F_4 = (2^n - 2^k)! \tag{11.6.4}$$

The product $F_1 F_2 F_3 F_4$ gives the number of permutations with exactly i input-output connections using the internal link considered. In general, there are $2^n!$ permutations. Assuming that all permutations are equally likely, the probability $p_{i,k}$ is a ratio of these two numbers, which can be simplified to give:

$$p_{i,k} = \frac{\binom{2^{n-k}}{i}\binom{2^n - 2^{n-k}}{2^k - i}}{\binom{2^n}{2^k}} \tag{11.6.5}$$

For $k > \frac{n}{2}$, we just substitute $n-k$ instead of k in the equation (11.6.5).

11.7 Exercises

1. Consider two queues in tandem, with exponentially distributed service times of mean $1/\mu_i$, $i=1,2$. Arrivals at the first queue is Poisson.

    a. Draw the two dimensional Markov chain for the states (k_1, k_2), where k_i is the number of customers in queue i.

    b. Show that the Markov chain is reversible using the notion of detail balance.

    c. Also, give the steady state probability for the Markov chain.

2. Now consider a closed cyclic queue formed from the two tandem queues in the previous problem by having the output of the second queue serves as the input to the first queue. There are L customers circulating in these queues.

    a. Draw the one dimensional Markov chain for k_1 ranging from 1 to L. (Implicitly, $k_2 = L - k_1$.)

    b. Find the steady state probabilities. Do they agree with the product form solution for closed networks?

    c. Is the Markov chain reversible?

    d. Explain why the arrival process to each queue is not Poisson, and why in the limit $L \to \infty$, the arrival process becomes Poisson.

3. Consider the following open cyclic network. We have an M/M/1 queue with arrival rate λ. A fraction f of customers, upon completion of service, leave the system for good. The remaining fraction $1-f$ of customers recirculates to the input of the M/M/1 queue, in addition to the arrivals with rate λ.

    a. Find the values of f for which the queue length can reach steady state.

11.7 Exercises

    b. Draw the Markov chain for the number of customers in the system.

    c. Show that the departure process of customers from the system is a Poisson process.

4. Let us study some properties of a reversed process.

    a. Show that the reversed process of a stationary Markov process is also a stationary Markov process. (Hint: Consider equation (7.2.2))

    b. Show that the transition probabilities of the reversed process is

$$\tilde{p}_{ss'} = \frac{p_{s's} p_{s'}}{p_s}$$

in which $p_{s's}$ is the transition probability for the forward process, and p_s and $p_{s'}$ are the steady state probabilities for the forward process.

5. Using the result of the previous problem, we shall show that at points of departure from the network, the exit processes are independent and Poisson.

    a. Draw this conclusion by arguing that the reversed process resembles the population distribution $(\tilde{k}_v)$ of a Jackson network with the same set of nodes but different arrival and departure processes.

    b. Describe the Jackson network for the reversed process. Specifically, show that the rate of arrivals to the network at a node v for the reversed process is

$$\tilde{\lambda}_v = g_v \alpha_v$$

Also, the rate of departures from the network at a node v is

$$\tilde{g}_v = \frac{\lambda_v}{\alpha_v}$$

The branching probabilities are given by

$$\tilde{f}_{v'v} = f_{vv'} \frac{\alpha_v}{\alpha_{v'}}$$

c. Using equation (11.2.2), show that $\tilde{\alpha}_v = \alpha_v$. Hence draw the conclusion that the population distributions for the forward and the reversed processes are identical using equation (11.2.3).

11.8 References

Product form solutions for steady state probabilities in queueing networks were developed in the papers by Burke, Jackson, and Baskett *et. al.* A detail treatment queueing networks as well as the notion of reversibility can be found in the book by Kelly. Buzen also developed methods for analyzing such networks. Applications of queueing network theory to computer networks can be found in the books by Kobayashi and by Kleinrock.

The delay and throughput analysis of banyan networks (with and without internal buffering) has been studied by Dias *et. al.*, by Patel and by Jeng. Our treatment simplifies that of Jeng. The study on the speed up required for buffered banyan network with load imbalance is new. Studies of the banyan network under permutation requests are also presented in the thesis by Szymanski.

1. F. Baskett, M. Chandy, R. Muntz, and J. Palacios, "Open, closed and mixed networks of queues with different classes of customers," *Journal of Associated Computing Machinery*, vol 22, pp. 248-260, 1975.

2. P. J. Burke, "Output processes and tandem queue," *Proc. Symp. on Computer-Communication and Teletraffic*, New York, pp. 419-429, April 1972.

3. J. P. Buzen, "Computational algorithms for closed queueing networks with exponential servers," *Communications of ACM*, vol. 16, pp. 527-531, 1973.

4. D. M. Dias and J. R. Jump, "Analysis and simulation of buffered delta networks," *IEEE Trans. on Computers*, vol. 30,

11.7 Exercises

no. 4, pp. 273-282, April 1981.

5. J. R. Jackson, "Networks of waiting lines," *Operations Research*, vol-5, pp. 699-704, August 1957.

6. Y. C. Jeng, "Performance analysis of a packet switch based on a single-buffered banyan network," *IEEE J. Selected Areas of Comm.*, vol-1, pp. 1014-1021, Dec. 83.

7. F. P. Kelly, *Reversibility and Stochastic Networks*, Wiley, Chichester, 1979.

8. L. Kleinrock, *Queueing Systems, Vol. 2: Computer Applications*, Wiley Interscience, New York, 1980.

9. H. Kobayashi, *Modeling and Analysis: An Introduction to System Performance Evaluation Methodology*, Addison-Wesley, Reading, MA, 1978.

10. J. H. Patel, "Performance of processor-memory interconnections for multiprocessor," *IEEE Trans. on Computers*, vol. 30, no. 10, pp. 771-780, Oct. 1981.

11. T. H. Szymanski, *On Interconnection Networks for Parallel Processors*, Ph. D. Dissertation, Department of Electrical Engineering, University of Toronto, Sept. 1987.

Index

A

Acyclic network 322
Aggregate traffic 187 212
Asymptotically non-blocking network 262-267
Alternating state renewal process 186
Alternative path routing 248
Appended switch 123
Arrival process 272
Asynchronous Transfer Mode (ATM) 25 31-35 179

B

Bandwidth allocation 237-238
Banyan network 99 126-137
Baseline network 104
Batcher sorting network 148-152
Bell system 16-18
Benes network 72 104
Bernoulli random variable 187
Binomial distribution 188
Birth-death process 216
Bitonic list 147
Bitonic sorter 147
Blocking 58 211
- Call 217
- System (or time) 216
Broadband 1
Broadband limit 199
Broadband service 9-11
Broadband technology 4-9

Buffered banyan network 119 321
Bulk arrivals 307
Burke's theorem 319
Bursty service 35-38

C

Campbell's theorem 198
Cantor network 74-77
Carried traffic 212
Cell (ATM) 31
Central limit theorem 201
Central office 18
Chernoff bound 202 233 241
Circuit switching 27 42 53
Clos network 61 250 256
Clos Theorem 64
Closed queueing network 323-326
Combinatorial bounds 57-62
Compact superconcentrator 56 62-67
Concentrator 56
Control algorithm 77-80
Convolution formula
- Poisson process 196
- Alternating state 192
Copy distribution network 102
Copy network 88 101-104
Crossbar 38
Crosspoint 4 38
Crosspoint complexity 57
Cyclic network 322

D

DS carriers 24
Data network 7
Datagram 48
Departure process 272
Depth-first-search circuit hunting 114-117
Detailed balance 219 316-318
Digital communications 6-7
Distributed Switching 7
Distribution network 99
Duplicated switch 123

E

Echo for telephony 48
Embedded Markov process 185 276
Engset formula 217
Equilibrium equation 185 253 324
Equivalent Random Method 228-232
Erlang 18
Erlang blocking formula 218
Erlang delay formula 275
Exchange network 3
Exponential distribution 183

F

Fan-out 57
Fast circuit switching 12
Fast packet switching 119-126
Filtered Poisson process 193-198
Finite state model 178-182
First Come First Served (FCFS) 120
Fixed length packets 31
Flip network 128
Frame- TDM 27
Frequency Division Multiplexing (FDM) 6

G

Gaussian distribution 201
General distribution 272
Generalized circuit switching 85
Geometric distribution 272 282
Geom/Geom/1 queue 282-284
Graph 54

H

Hayward approximation 229
Head of Line (HOL) blocking 142 159-161
Hierarchical network 3 17
Hundred Calling Seconds (CCS) 17
Hybrid TDM 34
Hypercube network 135

I

Improved large deviation approximation 202-207
Independent link assumption 254-256
Independent path assumption 254-256
Input buffering 120
Insensitivity of Blocking 221-227
Integrated access 12
Integrated call processing 13
Integrated network 11-14
Integrated switching 13
Integrated transport 13
Interarrival time distribution 272-273
Internal blocking 140
Interval splitting algorithm 165-168
Intraoffice calling percentage 17
ITT System-12 83

J

Jackson network 322

Index   345

Jacobaeus Method 256-262

K
Kleinrock's independence
  assumption 329

L
Label 31 47
Law of large numbers 199
Lee graph 249
Lee method 248-253
Linear complexity concentrator
  91 111
Little's result 275
Load imbalance
- packet switching 297-301
Local access network 2 17
Logical depth 57
Long distance network 3
Looping algorithm 78

M
M/G/1 queue 276-284
M/M/m queue 274-275
Markov inequality 199
Markov process 183
Measure function
- Poisson process 194
Memory switches 122
Moment generating function 190
- M/G/1 280-284
- Poisson process 298
- Alternating state 292
Merge-sort 146
Movable boundary TDM 29
Multi-layer traffic 50 179
Multi-media service 9
Multi-point circuit switching 54
Multi-point packet switching 164
Multi-point service 9
Multi-queue multi-server
  queueing system 272 284-290

Multi-rate services 9
Multi-slot TDM 29
Multi-stage packet network 313
Multi-stage switching network 60
Multi-stages of M/M/1 Queue 314
Multi-window TDM 28
Multiple arbitrations 159

N
Network control 58
Network management 59
Network symmetry assumption
  251 254
Non-series-parallel network 250
Number 5 crossbar switch 108
Number 4 ESS switch 83
Number 5 ESS switch 83

O
OTCi carriers 49
Offered traffic 212
Open queueing network 322-326
Output buffering 121
Output conflict 46 141
Output conflict resolution
  141 152-158
Output unspecific interconnection
  network 86
Overflow traffic 212 228-232

P
Packet switching 27 46
Packetization delay 32 48
Parallel network 250
Path 54
Paull's matrix 61
Paull's theorem 69
Peakedness 228
Pippenger's multi-point switch
  106-108
Point-to-point circuit switching
  53

Poisson distribution 188-189
Poisson process 188-191 193-198
Pollaczek-Khinchin
- mean-value formula 280 305
- transform formula 282
Product form solution
- Circuit switching 222
- Packet switching 324
Public Switched Telephone
  Network (PSTN) 16

Q
Queueing
- Buffered banyan network 331-334
- multi-cast network 301-304
Queueing networks 322-326

R
Randomized routing 132
Re-entry packet switch 154-155
Rearrangeable network 63
Recursive construction 70-74
Reversed process 315
Reversibility 315-318
Row major assignment
  algorithm 96
Running sum adder network 165

S
SONET 50
SS switching 44
SSS switching 44
Saturation throughput
- Input queueing 296
- Multi-casting 304
- Unbalanced load 301
Scheduler 26
Self-routing multi-cast
  switch 164-170
Self-routing network 126-131
Semi-Markov process 183
Series-parallel network 250-253

Service time distribution 272-273
Shear sort 171
Shuffle exchange network 126-137
Signal sampling 6
Single slot TDM 27
Single-stage packet network 271
Single-stage resource sharing 211
Slepian-Duguid theorem 66
Slot matching 65
Sorting network 143
Space Division Multiplexing
  (SDM) 38
Space Division Switching
  (SDS) 6 38-40
Speech traffic 178-181
State transition 182-184
Stationary process 182
Steady state probabilities 184-186
Step-by-step switching 4 115
Sterling's approximation 90
Strict-sense non-blocking
  network 58 63-66
Superconcentrator 56
Superposition of traffic 182-185
Synchronous Transfer Mode
  (STM) 27-31

T
T1 carrier 49
TS switching 44
TST switching 44
Tail distribution 201
Tandem queues 314
Terminal model 178-184
Terminal symmetry assumption
  251 254
Three phase algorithm 155-158
Three stage factorization 70
Time Division Multiplexing
  (TDM) 6-7 26-27
Time Division Switching
  (TDS) 7

Index

Time Multiplexed Switch
  (TMS) 42-46
Time Slot Interchange
  (TSI) 39
Time slot mismatch 117-119
Token 152
Traffic engineering 232
Truncated Markov chain 216 221
Truncated Poisson distribution 214
Truncated binomial distribution 214
Truncated random variable 213
Trunk 3
Trunk group 3
Two stage factorization 94

U
Unique crossover property 148
Unique path routing 247

V
Variable length packet 33 47
Video traffic 181-182 208
Virtual Circuit Identifier (VCI) 47
Virtual bandwidth 238
Virtual circuit 47
Virtual cut through 121

W
Wide-sense non-blocking
  network 59
Window 29

Z
Z-transform 280
Zero-one principle 170-171